INTEGRATED PEST MANAGEMENT FOR

POTATOES

IN THE WESTERN UNITED STATES

INTEGRATED PEST MANAGEMENT FOR

POTATOES

IN THE WESTERN UNITED STATES

SECOND EDITION

WESTERN REGIONAL IPM PROJECT
COLORADO STATE UNIVERSITY
UNIVERSITY OF IDAHO
MONTANA STATE UNIVERSITY
OREGON STATE UNIVERSITY
WASHINGTON STATE UNIVERSITY

UNIVERSITY OF CALIFORNIA
STATEWIDE INTEGRATED PEST MANAGEMENT PROGRAM

AGRICULTURE AND NATURAL RESOURCES
PUBLICATION 3316

2006

iv

PRECAUTIONS FOR USING PESTICIDES

Pesticides are poisonous and must be used with caution. READ THE LABEL BEFORE OPENING A PESTICIDE CONTAINER. Follow all label precautions and directions, including requirements for protective equipment. Use a pesticide only on the plants or site specified on the label or in published University of California recommendations. Apply pesticides at the rates specified on the label or at lower rates if suggested in this publication. In California, all agricultural uses of pesticides must be reported, including use in many non-farm situations, such as cemeteries, golf courses, parks, roadsides, and commercial plant production including nurseries. Contact your county agricultural commissioner for further details. Laws, regulations, and information concerning pesticides change frequently, so be sure the publication you are using is up-to-date.

Legal Responsibility. The user is legally responsible for any damage due to misuse of pesticides. Responsibility extends to effects caused by drift, runoff, or residues.

Transportation. Do not ship or carry pesticides together with food or feed in a way that allows contamination of the edible items. Never transport pesticides in a closed passenger vehicle or in a closed cab.

Storage. Keep pesticides in original containers until used. Store them in a locked cabinet, building, or fenced area where they are not accessible to children, unauthorized persons, pets, or livestock. DO NOT store pesticides with foods, feed, fertilizers, or other materials that may become contaminated by the pesticides.

Container Disposal. Consult the pesticide label, the County Department of Agriculture, or the local waste disposal authorities for instructions on disposing of pesticide containers. Dispose of empty containers carefully. Never reuse them. Make sure empty containers are not accessible to children or animals. Never dispose of containers where they may contaminate water supplies or natural waterways. Offer empty containers for recycling if available. Home use pesticide containers can be thrown in the trash only if they are completely empty.

Protection of Nonpest Animals and Plants. Many pesticides are toxic to useful or desirable animals, including honey bees, natural enemies, fish, domestic animals, and birds. Certain rodenticides may pose a special hazard to animals that eat poisoned rodents. Plants may also be damaged by misapplied pesticides. Take precautions to protect nonpest species from direct exposure to pesticides and from contamination due to drift, runoff, or residues.

Permit Requirements. Certain pesticides require a permit from the county agricultural commissioner before possession or use.

Plant Injury. Certain chemicals may cause injury to plants (phytotoxicity) under certain conditions. Always consult the label for limitations. Before applying any pesticide, take into account the stage of plant development, the soil type and condition, the temperature, moisture, and wind. Injury may also result from the use of incompatible materials.

Personal Safety. Follow label directions carefully. Avoid splashing, spilling, leaks, spray drift, and contamination of clothing. NEVER eat, smoke, drink, or chew while using pesticides. Provide for emergency medical care IN ADVANCE as required by regulation.

Worker Protection Standards. Federal Worker Protection Standards require pesticide safety training for all employees working in agricultural fields, greenhouses, and nurseries that have been treated with pesticides, including pesticide training for employees who don't work directly with pesticides.

ORDERING

For information about ordering this publication and/or a free catalog, contact
University of California
Agriculture and Natural Resources
Communication Services
6701 San Pablo Avenue, 2nd Floor
Oakland, California 94608-1239
Telephone 1-800-994-8849
(510) 642-2431
FAX (510) 643-5470
E-mail: danrcs@ucdavis.edu
Visit the ANR Communication Services Web site at
http://anrcatalog.ucdavis.edu

Publication 3316

Other books in this series include:
Integrated Pest Management for Alfalfa Hay, Publication 3312
Integrated Pest Management for Almonds, Second Edition, Publication 3308
Integrated Pest Management for Apples and Pears, Second Edition, Publication 3340
Integrated Pest Management for Citrus, Second Edition, Publication 3303
Integrated Pest Management for Cole Crops and Lettuce, Publication 3307
Integrated Pest Management for Cotton, Second Edition, Publication 3305
Integrated Pest Management for Floriculture and Nurseries, Publication 3402
Integrated Pest Management for Rice, Second Edition, Publication 3280
Integrated Pest Management for Small Grains, Publication 3333
Integrated Pest Management for Stone Fruits, Publication 3389
Integrated Pest Management for Strawberries, Publication 3351
Integrated Pest Management for Tomatoes, Fourth Edition, Publication 3274
Integrated Pest Management for Walnuts, Third Edition, Publication 3270
Natural Enemies Handbook: The Illustrated Guide to Biological Pest Control, Publication 3386
Pests of Landscape Trees and Shrubs, Second Edition, Publication 3359
Pests of the Garden and Small Farm, Second Edition, Publication 3332

ISBN-13: 978-1-879906-77-8
ISBN-10: 1-879906-77-5
Library of Congress Control Number: 2005909526
©1992, 2006 by the Regents of the University of California
Division of Agriculture and Natural Resources
All rights reserved.
No part of this publication may be reproduced, stored in a retrieval system, or transmitted, in any form or by any means, electronic, mechanical, photocopying, recording, or otherwise, without the written permission of the publisher and the authors.

Printed in Canada on recycled paper.

 This publication has been anonymously peer reviewed for technical accuracy by University of California scientists and other qualified professionals. This review process was managed by the ANR Associate Editor for Pest Management.

The University of California prohibits discrimination against or harassment of any person employed by or seeking employment with the University on the basis of race, color, national origin, religion, sex, physical or mental disability, medical condition (cancer-related or genetic characteristics), ancestry, marital status, age, sexual orientation, citizenship, or status as a covered veteran (special disabled veteran, Vietnam-era veteran or any other veteran who served on active duty during a war or in a campaign or expedition for which a campaign badge has been authorized). University Policy is intended to be consistent with the provisions of applicable State and Federal laws. Inquiries regarding the University's nondiscrimination policies may be directed to the Affirmative Action/Staff Personnel Services Director, University of California, Agriculture and Natural Resources, 300 Lakeside Drive, 6th floor, Oakland, CA 94612-3550; (510) 987-0096. **For information about ordering this publication, telephone 1-800-994-8849.**

To simplify information, trade names of products have been used. No endorsement of named or illustrated products is intended, nor is criticism implied of similar products that are not mentioned or illustrated.
3m-pr-1/06-SB/CR

Contributors and Acknowledgments

Second edition revisions
written by Larry L. Strand
First edition written by Larry L. Strand and
Paul A. Rude

Photographs by Jack Kelly Clark
(except as noted)

Mary Louise Flint, Technical Editor

Prepared by IPM Education and Publications, an office of the University of California Statewide IPM Program at Davis.

This manual was produced under the auspices of the University of California Statewide IPM Program with additional funding from the California Potato Advisory Board. Assistance and additional funding for the first edition was provided by the SAES Western Regional Research Project W-161; the University of Idaho; Washington, Colorado, Oregon, and Montana State Universities; and the California Potato Advisory Board.

Technical Coordinators for Second Edition

Rick A. Boydston, Research Plant Physiologist, USDA-ARS, Irrigated Agriculture Research and Extension Center, Prosser, WA

Philip B. Hamm, Extension Plant Pathologist, Oregon State University, Hermiston Agricultural Research and Extension Center, Hermiston

Russell E. Ingham, Professor, Oregon State University, Corvallis

Stephen L. Love, Professor, University of Idaho, Aberdeen Research and Extension Center, Aberdeen

Alvin R. Mosley, Professor and Potato Specialist, Oregon State University, Corvallis

Thomas M. Mowry, Associate Professor, University of Idaho, Parma Research and Extension Center, Parma

Nora L. Olsen, Associate Professor and Extension Potato Specialist, University of Idaho, Twin Falls Research and Extension Center, Twin Falls

Ronald E. Voss, Extension Vegetable Specialist, University of California, Davis

Contributors to Second Edition

Abbreviations after contributors' names stand for these institutions: (ARS)—United States Department of Agriculture, Agricultural Research Service; (CS)—Colorado State University; (FWS)—United States Fish and Wildlife Service; (MS)—Montana State University; (OS)—Oregon State University; (UC)—University of California; (UI)—University of Idaho; (WS)—Washington State University

Entomology: Sue L. Blodgett (MS), Whitney Cranshaw (CS), Thomas M. Mowry (UI)

Horticulture, Physiology: Dan C. Hane (OS), Steven R. James (OS), Stephen L. Love (UI), Jeffrey P. McMorran (OS), Alvin R. Mosley (OS), Nora Olsen (UI), Robert E. Thornton (WS), Ronald E. Voss (UC), Dale T. Westermann (ARS)

Nematology: Saad L. Hafez (UI), Russell E. Ingham (OS), Hassan Mojtahedi (ARS)

Plant Pathology: R. Michael Davis (UC), Philip B. Hamm (OS), Dennis A. Johnson (WS), Cynthia M. Ocamb (OS), Mary L. Powelson (OS), Jonathan L. Whitworth (ARS)

Vertebrates: Marco Buske (FWS), Rex E. Marsh (UC)

Weed Science: Rick A. Boydston (ARS), Jed B. Colquhoun (OS), Pamela J. S. Hutchinson (UI), Corey V. Ransom (OS)

Special Thanks

The following have generously provided information, offered suggestions, reviewed draft manuscripts, or helped obtain photographs: H. L. Carlson, L. L. Ewing, D. Haviland, D. Inglis, E. Natwick, P. Nolte, J. Nunez, M. K. O'Neill, G. Q. Pelter, K. A. Rykbost, C. C. Shock, G. A. Secor, P. E. Thomas, S. Yilmaz.

We would also like to acknowledge the important role of the contributors to the first edition of this manual, which was published in 1986 and 1992: H. S. Agamalian, T. C. Allen, A. Appleby, R. Arnold, O. Bacon, R. G. Beaver, R. Benton, P. Bessey, G. W. Bishop, T. Bowman, C. Burkhardt, V. E. Burton, R. L. Chase, D. Davis, J. R. Davis, G. D. Easton, L. Fox, G. D. Franc, B. Gillespie, J. Guerard, L. C. Haderlie, D. H. Hall, M. D. Harrison, W. Hart, E. Heikes, L. K. Hiller, H. Homan, W. M. Iritani, H. J. Jensen, J. Kelley, H. M. Kempen, G. E. Kleinkopf, G. D. Kleinschmidt, J. Knight, K. W. Knutson, P. A. Koepsell, C. Livingston, G. A. McIntyre, E. L. Nigh, A. G. Ogg, Jr., R. Parker, J. J. Pavek, J. Pinkerton, D. Powell, P. A. Roberts, G. S. Santo, C. C. Stanger, M. E. Stanghellini, M. K. C. Sun, I. J. Thomason, H. Timm, H. Toba, K. B. Tyler, A. R. Weinhold

Production

Design (second edition): Celeste Rusconi, ANR Communication Services

Design (IPM manual series): Seventeenth Street Studios

Drawings (second edition): Celeste Rusconi

Drawings (first edition): Marvin Ehrlich, Franz Baumhackl

Editing: Stephen Barnett, ANR Communication Services

Contents

Integrated Pest Management for Potatoes in the Western United States

This manual is designed to help growers, crop consultants, pest control professionals, and farm managers apply principles of integrated pest management (IPM) to potato crops in the western United States.

Integrated pest management is an ecosystem-based strategy that focuses on long-term prevention of pests or their damage through a combination of techniques such as biological control, habitat modification, modification of cultural practices, and use of resistant varieties. Pesticides are used only when needed and according to established guidelines, and treatments are made with the goal of removing only the target organism. Pest control materials are selected and applied in a manner that minimizes risks to human health, beneficial and nontarget organisms, and the environment.

In an IPM program, pest management is coordinated with production practices, and emphasis is placed on anticipating and preventing problems whenever possible. The term "pest" is used in this manual to include destructive insects and other arthropods, pathogens, nematodes, weeds, and vertebrates.

The second chapter of this manual summarizes crop growth and development. It provides a background for understanding how potato growth and production are influenced by pest injury, cultural practices, and other factors. The discussion of management methods in the third chapter places integrated pest management in perspective with other production practices. Later sections on specific pests stress biological information that relates to management strategies and the pest's impact on the crop. The photos and descriptions help to identify pests and the symptoms of pest damage. Charts and forms are presented to help organize pest management activities. The glossary includes technical terms and other terminology that may be unfamiliar to readers.

Pesticides are discussed where appropriate, but because pesticide registrations often change, this manual does not make specific recommendations. However, the suggested reading at the end of the manual lists sources of current recommendations from universities in the western states. Research is continuing in many areas of potato pest management, so monitoring and control methods may change as new information and crop cultivars become available. Stay

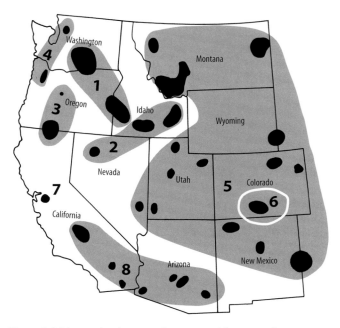

Figure 1. Major production areas for commercial potatoes (potatoes not grown for seed) in the western United States: (1) Columbia Basin and lower Snake River Valley; (2) Upper Snake River Valley and northern Nevada; (3) Central Oregon and Klamath Basin; (4) Northern coastal valleys; (5) Rocky Mountain valleys and plains; (6) San Luis Valley; (7) Sacramento–San Joaquin Delta; and (8) Low-elevation desert valleys.

Figure 2. Major areas of seed potato production in the western United States.

up to date on pest management methods by keeping in touch with local cooperative extension agents or other experts.

The western states (Figs. 1 and 2) produce nearly two-thirds of the potatoes grown in the United States (Table 1). The majority of acreage is planted to russet cultivars such as Russet Burbank and Russet Norkotah. Over 15 other cultivars are also grown in one or more areas of the region. In most of the growing areas potatoes are planted in the spring and harvested in late summer and fall. An exception is the low desert valley region (area 8, Fig. 1), where most of the acreage is planted in fall or winter and harvested in late spring or early summer. The growing season in most areas ranges between 120 and 140 days. However, some high-elevation areas have 90- to 100-day seasons, while some areas in the Columbia Basin have 160- to 180-day seasons.

A number of pests are important to all western potato-growing areas. Aphids, primarily the green peach aphid and the potato aphid, are vectors of leafroll and other potato viruses, important diseases wherever potatoes are grown. Verticillium wilt, late blight, and blackleg can be problems everywhere; bacterial ring rot can occur, but its incidence is minimized by the use of certified seed. Rhizoctonia stem and stolon canker is always a threat if sprout growth is delayed or occurs in cool, wet soil. In all growing areas nightshades, pigweeds, lambsquarters, and annual grasses are problems, and nutsedge is a potential problem.

Other pests are important only in certain parts of the western region. Root knot nematodes occur wherever potatoes are grown, but they are most serious in Pacific Northwest growing locations, areas 1 to 3 of Figure 1. Potato psyllid occurs throughout the West; it is a major problem in the Rocky Mountain growing locations, areas 5 and 6, and in some low-elevation desert valley locations, area 8. Quackgrass and Canada thistle are problem weeds in most of the Pacific Northwest and Rocky Mountain growing locations, areas 1 to 6. Field bindweed occurs occasionally in all areas; it is more frequently a pest in parts of areas 1, 2, and 5. Kochia, cocklebur, and sunflower are common weeds in areas 1, 2, 3, 5, and 6.

Russet Burbank, Ranger Russet, and Russet Norkotah are the main cultivars grown in the Columbia Basin and lower Snake River Valley, area 1. Other cultivars include Shepody, Alturas, Umatilla Russet, Yukon Gold, and Atlantic. Planting begins in March and April. Harvests of early cultivars such as Russet Norkotah begin in July; late cultivars such as Russet Burbank are harvested from late August through early November. The Columbia Basin of Washington and Oregon has the longest growing season, up to 180 days, with late harvest yields of 500 to 850 hundredweight per acre (56.0 to 92.3 t/ha), the highest reported yields in the world. Yields in the lower Snake River Valley average 350 to 500 hundredweight per acre (39.2 to 56.0 t/ha). Important insect pests in these areas include aphids, Colorado potato beetle, and wireworms. The potato tuberworm recently has become a problem in the Columbia Basin. Leafroll and net necrosis, other potato viruses, Verticillium wilt, late blight, blackleg, Rhizoctonia, and silver scurf on fresh market potatoes are major disease problems; powdery scab, powdery mildew, and Sclerotinia are occasional problems. Root knot nematodes and stubby-root nematode, the vector of corky ringspot, are serious pests. Important weeds include Canada thistle, quackgrass, field bindweed, nutsedge, dodder, nightshades, pigweeds, lambsquarters, kochia, cocklebur, and sunflower.

The majority of acreage in the upper Snake River Valley and northern Nevada, area 2, is planted to Russet Burbank; other cultivars grown include Ranger Russet, Russet Norkotah, Gem Russet, Alturas, Bannock Russet, Cal White, and Shepody. Planting takes place in April and harvest in September and October, with yields averaging 280 to 400 hundredweight per acre (31.4 to 44.8 t/ha). Aphids, Colorado potato beetle, wireworms, Verticillium wilt, late blight, blackleg, Rhizoctonia, leafroll and other potato viruses, root knot nematodes, and early blight are important pests in this area. Problem weeds include Canada thistle, quackgrass, nightshades, pigweeds, lambsquarters, kochia, cocklebur, and sunflower.

Central Oregon and the Klamath Basin of southern Oregon and northern California, area 3, produce potatoes for seed and for fresh market. The most common cultivars are Russet Norkotah, Russet Burbank, proprietary Frito-Lay and other chipping cultivars, and a small but increasing acreage of red-skinned and yellow-flesh cultivars. Planting is in May and harvest in September and October, with yields averaging 375 to 400 hundredweight per acre (42.0 to 44.8 t/ha). Frost is a threat both early and late in the season; solid set sprinklers can protect the foliage from frost damage. Colorado potato beetle is a pest in central Oregon but not in the Klamath Basin; aphids, wireworms, Verticillium wilt, late

blight, blackleg, potato viruses, Rhizoctonia, white mold, silver scurf, and root knot nematodes are problems in both areas. Early blight and pink rot are important pests in the Klamath Basin, where corky ringspot is sometimes a problem in coarse, sandy soils. In some years, voles (meadow mice) become serious pests in parts of the Klamath Basin. Major weed pests include Canada thistle, quackgrass, nightshades, kochia, and wild oats. Several mustard family weeds are important pests, primarily because they can be early-season hosts for green peach aphids.

The relatively cool, damp climate of the northern coastal valleys, area 4, makes late blight a perennial problem. The possibility of wet weather after planting always makes seed piece decay a potential threat. Potato acreage in the Willamette Valley of Oregon is planted to chipping cultivars, primarily Snowden, Atlantic, and proprietary Frito-Lay cultivars, and to Russet Burbank, Russet Norkotah, and red-skinned cultivars for fresh market. Planting is in May and harvest from mid-July to early October. Yields average 275 to 375 hundredweight per acre (30.8 to 42.0 t/ha), depending on harvest date and cultivar. The main cultivars grown in northwest Washington are Russet Norkotah, chipping, and red-skinned cultivars. A range of cultivars is grown for seed. Planting is in May and harvest from September through November. Yields of commercial plantings are about 325 to 375 hundredweight per acre (36.4 to 42.0 t/ha). Late blight is the major pest problem; aphids, the viruses they transmit, and Rhizoctonia are also important. Quackgrass and annual grasses are the major weed problems.

The Rocky Mountain valleys and plains, area 5, have seasons and yields similar to the upper Snake River Valley. Cultivars grown include Russet Norkotah, Russet Nugget, Atlantic, Yukon Gold, Sangre, and a number of others. Colorado potato beetle occurs throughout this region, and it is a serious pest in the most northern areas. Potato psyllid is

Table 1. Potato Acreage (Thousands of Acres Harvested) and Production (Thousands of Hundredweight [cwt]) in the Western United States.

	1999		2000		2001		2002		2003		2004	
	Acres	Cwt	Acres	Cwt	Acres	Cwt	Acres	Cwt	Acres	Cwt	Acres	Cwt
Arizona	9.6	3,024	9.0	2,520	8.2	2,214	7.8	2,106	7.6	2,090	6.2	1,767
California	43.2	16,227	44.0	16,710	35.9	13,188	43.6	17,069	43.1	17,139	45.1	18,099
Colorado	84.4	28,237	83.9	30,960	73.4	23,373	77.8	30,153	72.4	26,198	70.7	25,484
Idaho	393.0	139,330	413.0	152,320	348.0	120,200	373.0	133,385	358.0	123,180	353.0	131,970
Montana	10.9	3,325	11.3	3,503	10.3	3,296	10.4	3,224	10.6	3,339	10.6	3,551
Nevada	6.5	2,860	7.0	3,150	6.5	2,340	7.6	2,660	8.0	3,320	6.7	2,881
New Mexico	10.9	3,755	9.8	3,770	6.4	2,198	6.3	2,336	5.9	2,132	5.0	2,060
Oregon	55.5	28,020	56.5	30,683	44.5	20,730	49.8	24,936	42.6	20,991	37.0	19,775
Utah	2.0	580	1.5	435	1.3	345	0.8	244	1.0	335	—	—
Washington	170.0	95,200	175.0	105,000	160.0	94,400	162.0	92,340	162.0	93,150	159.0	93,810
Wyoming	0.5	148	—	—	—	—	—	—	—	—	—	—
Total	786.5	320,706	811.0	349,051	694.5	282,284	739.1	308,453	711.2	291,874	693.3	299,397
U.S. total	1,331.8	478,093	1,347.5	513,544	1,220.9	437,673	1,265.9	458,171	1,250.3	459,045	1,168.1	456,362

Metric conversions: 1 acre = 0.4047 ha; 1 cwt = 45.36 kg.
Source: Data obtained from USDA National Agricultural Statistics Service (www.usda.gov/nass).

a sporadic and occasionally serious problem. Early blight can be a major disease problem in most of the Rocky Mountain areas. Verticillium wilt, blackleg, and potato viruses are also important diseases. Root knot nematodes are occasionally serious in some areas. Quackgrass is a problem in some areas; Canada thistle is important throughout the region and growing in severity in central and northwestern New Mexico. Other problem weeds include nutsedge, field bindweed, nightshades, pigweeds, lambsquarters, and annual grasses.

The San Luis Valley, area 6, is at a higher elevation than the other Rocky Mountain valleys. A large proportion of the acreage is planted for seed production. The main cultivar grown for commercial production is Russet Norkotah; Russet Nugget, Centennial Russet, Yukon Gold, and small amounts of several other cultivars are also grown. Planting is in late spring, harvest in early fall, and yields average about 300 hundredweight per acre (33.6 t/ha). Pest problems are similar to the other Rocky Mountain areas; psyllids are occasionally serious, but Colorado potato beetle is not a problem. Early blight, both foliar blight and tuber rot, is a serious disease, as well as Verticillium wilt and blackleg. Powdery scab occurs on some production sites. Aphids and viruses transmitted by them are important. Important weed pests include Canada thistle, nutsedge, nightshades, pigweeds, lambsquarters, and annual grasses.

Several potato cultivars are grown in the Sacramento–San Joaquin Delta of central California, area 7, and yields average 300 to 400 hundredweight per acre (33.6 to 44.8 t/ha). Summer crops of primarily Cal White, White Rose, Yukon Gold, and Chieftain are grown, and seed potatoes are planted in the summer for winter harvest. As in all other growing areas, aphids and the potato viruses they transmit are important pests. Nematodes generally are not a problem in the peat soils of the Delta; however, the high organic matter content increases the potential for common scab. Late blight occurs to some extent every season, and blackleg, Rhizoctonia, and Verticillium wilt are always potential problems. Important weed pests in these areas include nightshades, pigweeds, mallow, nettle, annual sowthistle, and mustards.

In the low-elevation desert valleys of Arizona and California, area 8, most potatoes are planted from November to February and harvested from May to August. The winter weeds nettle and London rocket may be problems in these areas. Black nightshade is the most prevalent nightshade species; mallow, purslane, and Russian thistle are also common weed pests. In California, some plantings are made in the summer and harvested in the winter using ground storage. Cal White, White Rose, Russet Norkotah, Atlantic, Chipeta, Red LaSoda, Yukon Gold, and Cherry Red are the major cultivars grown in California, with yields of 300 to 500 hundredweight per acre (33.6 to 56.0 t/ha). Pest problems include tuberworm, wireworms, psyllids, common scab, powdery scab, silver scurf, early dying, Rhizoctonia, white mold, Sclerotium stem rot, and late blight. Root knot nematodes are occasionally serious when tubers are harvested in July or August. In Arizona the main cultivars are Atlantic, Russet Norkotah, and Red LaSoda; yields average 280 to 300 hundredweight per acre (31.4 to 33.6 t/ha). Major pest problems include potato psyllid, blackleg, and Pythium tuber rot.

Most seed-producing areas are high-elevation, cool, short-season areas (see Fig. 2). Location, climate, and the certification requirements for fields and seed tubers used for planting help reduce the severity of pest problems in these growing areas. Green peach aphid is the most important pest because it transmits several potato viruses. Control measures are designed to prevent any aphid buildup in seed fields, and in some cases to prevent their development on alternate hosts or in commercial potato fields from which they can transmit viruses to seed fields. Colorado potato beetle is a problem in Idaho, Wyoming, and some of the areas in Montana and Oregon. Cool weather and the finer-textured soils of many seed areas help keep root knot nematode populations from building up; certification requirements usually restrict the growth of certified seed potatoes to fields that have been free of root knot symptoms. Common weed pests include nightshades, pigweeds, lambsquarters, Canada thistle, and quackgrass. Control of mustard family weeds is important because they can be hosts for early-season development of green peach aphid populations.

Growth and Development Requirements of the Potato Plant

To accurately evaluate pest injury and distinguish it from other kinds of plant stress, it is important to understand the biology of the crop. IPM methods are chosen for their effects on the crop as a system, not just for their abilities to control specific pests. Successful pest management in potatoes requires that you know how pests and management practices interact to affect potato growth and development.

Growth Requirements

The potato plant, like a factory, requires a supply of raw materials and a source of energy to make a product. The raw materials for plant growth and tuber production are carbon dioxide, oxygen, water, and soluble mineral nutrients. Like other green plants, potatoes use complex chemical structures including chlorophyll, the green pigment of plants, to capture energy from sunlight. Some of the trapped energy is used to convert water and carbon dioxide from the air into sugars and other compounds that serve as the plant's energy supply and as building blocks for growth. This process is called photosynthesis.

Most water absorbed from the soil by plants passes from the roots up through the vascular system and evaporates through leaf pores called stomata (or "stomates"). Only a small amount of the water taken up by plants actually remains in their tissues. Most evaporation from leaves, which is called transpiration, occurs because stomata remain open to expose a moist surface that can absorb carbon dioxide from the air. The flow of water from roots to leaves also supplies water used in photosynthesis, carries nutrients throughout the plant, provides rigidity through plant cell turgor, and serves as a means of cooling. Higher temperature, lower humidity, and stronger wind generally increase transpiration and, therefore, the amount of soil moisture withdrawn by plants. However, under extremely hot, dry conditions, stomata close, stopping transpiration and evaporative cooling.

Plants absorb soluble mineral nutrients from the soil to build proteins, chlorophyll, and other components. Nutrients needed in the largest amounts are nitrogen, phosphorus, potassium, calcium, magnesium, and sulfur. Nutrients required only in very small amounts but still essential are iron, boron, manganese, zinc, molybdenum, copper, and chlorine.

Respiration is the process by which sugars and other compounds are metabolized to provide the energy that drives many growth processes. The byproducts of respiration are heat, water, and carbon dioxide. The availability of oxygen is critical for the respiration of roots and tubers as well as for aboveground plant parts. Both roots and tubers become more susceptible to certain diseases or physiological disorders in waterlogged or poorly aerated soil.

Development

In nature, the potato is a perennial plant that survives from year to year as a tuber, which is a modified underground stem. When grown as an annual crop, potatoes are propagated vegetatively through the planting of tubers or parts of tubers (seed pieces) with associated buds or "eyes" rather than true seeds. Potato plants do produce true seeds, but in the United States these are used for propagation primarily by plant breeders developing new cultivars.

A number of diseases can be transmitted in potato tubers. Potato plants that grow from tubers left in fields or cull piles are important sources of inoculum for these diseases. Foundation seed and seed certification programs have been developed to obtain and maintain sources of seed tubers that are relatively free from disease.

Growth and development of potato plants can be divided into five stages (Fig. 3):
- Sprout development (stage I)
- Vegetative growth (stage II)
- Tuber initiation (stage III)
- Tuber growth (stage IV)
- Maturation (stage V)

The timing of the events that make up this pattern of growth is influenced by cultivar, age of seed pieces, weather, cultural practices, pest damage, and other factors.

Stage I: Sprout Development

The potato tuber is a modified stem and its eyes are clusters of modified buds (Fig. 4). The rudimentary leaf scar below each bud cluster resembles an eyebrow, thus the name "eyes." Either whole tubers or pieces of tubers, each with one or more eyes, are planted. The emergence of sprouts from eyes is favored by dark, warm, moist conditions. Sprouts grow upward to emerge from the soil and form the stems of the plant. Buds form at swollen nodes along the stem.

Before emergence, seed pieces may decay, and the decay organisms may invade the base of the stem, weakening the plant and reducing yields. Developing stems are susceptible to certain other diseases, such as Rhizoctonia and blackleg. Vigorous sprout growth is more resistant to infection. Generally, the more rapidly sprouts emerge, the less susceptible they are to damage. The rate of sprout growth and, consequently, the time until emergence are temperature-dependent (Table 2), and therefore somewhat dependent on soil type and planting depth.

Starch in the seed piece supplies energy for sprout growth and development. There is usually enough starch in a sound and properly sized potato seed piece to support plant growth for 30 days or longer. About half of the starch is used up by the time the plant starts to produce sufficient energy to support further growth and development and accumulate reserves for tuber production. Food reserves and mineral nutrients continue to be used until they are exhausted or the seed piece is decayed. Inorganic nutrients, some of which must be supplied as preplant fertilizer materials, are generally necessary for most active early growth.

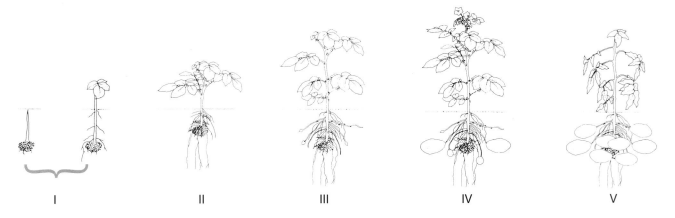

| I | II | III | IV | V |

Figure 3. Growth stages of the potato plant. I *Sprout Development:* Sprouts develop from eyes on seed tuber and grow upward to emerge; roots begin to develop. II *Vegetative Growth:* Branch stems and first 8 to 12 leaves develop from aboveground nodes along emerged sprouts; roots and stolons develop from belowground nodes. III *Tuber Initiation:* Tubers begin to form at tips of stolons; foliage continues to develop. IV *Tuber Growth:* Most of plant production supports tuber enlargement. V *Maturation:* Tuber periderm (skin) thickens, and dry matter content reaches a maximum; vines begin to senesce.

Roots develop from the base of each sprout, from underground nodes, often from stolons, and occasionally from tubers (see Fig. 4). The root system that develops is highly branched. Most of the nutrients that move up through the vascular system (Fig. 5) are absorbed by fine root hairs at the tips of root branches. By the time sprouts emerge, there is an extensive root system above the level of the seed piece. A fully developed plant has a root system that is concentrated in the top 18 inches (45 cm) of soil, although in some cases roots may extend as deep as 5 feet (150 cm) depending on moisture availability and the presence of restrictive soil layers.

Stage II: Vegetative Growth

Usually one to three sprouts from each seed piece develop into the mainstems of the plant. Leaves and branch stems develop from stem nodes after the stems emerge from the soil. The potato leaf is compound, consisting of a petiole with a terminal leaflet, two to four pairs of primary leaflets, and secondary leaflets interspersed along the petiole (see Fig. 4). Leaflet characteristics depend on cultivar. Once leaves are formed, a potato plant is able to produce all the energy it needs for further growth and development.

Stolons are lateral shoots that develop from buds at the underground nodes of stems. The vascular system of a stolon has a large amount of phloem tissue for rapid translocation of nutrients into growing tubers (see Fig. 5). Tubers develop just behind the tips of stolons. If stolons emerge from the soil, they form leafy shoots called secondary stems. Stolons produced on secondary stems do not always form tubers.

Depending on their growth habit (pattern of vine growth), potato cultivars may be classified as determinate or indeterminate. Late-season cultivars such as Russet Burbank and Ranger Russet are indeterminate. They have a vining type of growth habit, forming a succession of branch stems that become the main growing point and produce new foliage and flowers. Production of the first flowers (primary flower-

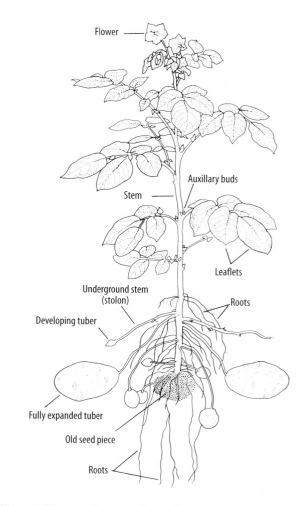

Figure 4. Diagram of a potato plant. Adapted from *Commercial Potato Production in North America, Potato Association of America Handbook,* Potato Association of America, Orono, ME, 1980.

ing) approximately coincides with the beginning of the tuber growth phase and occurs after the mainstem has formed 8 to 12 nodes. A branch then develops from an axillary bud one or two nodes below the flowering tip, producing 8 to 12 nodes and then secondary flowers. After secondary flowering, another branch develops in a similar fashion from below the secondary flowers; its growth terminates with a tertiary flowering. Indeterminate cultivars continue producing new foliage and in some cases may keep initiating new tubers until disease, weather conditions, or diminishing nutrients or water stop their growth. Completion of their growth cycle requires 100 or more days.

Early-season cultivars such as Russet Norkotah and Red Norland have a determinate, bush type of growth habit. Branches may develop after primary flowering, but always from nodes on the main stem. Tuber initiation and the onset

Table 2. Approximate Times for Russet Burbank Sprouts to Emerge When Planted at Different Soil Temperatures and Depths. Emergence times are affected by cultivar, conditions of seed storage, and soil physical conditions.

Soil temperature at planting	Days to sprout emergence at planting depth	
°F (°C)	4 inches (10 cm)	6 inches (15 cm)
40 (4.4)	40+	40+
45 (7.2)	30	40
50 (10.0)	25	28
55 (12.8)	22	26
60 (15.6)	22	24
65 (18.3)	20	22

of tuber growth occur earlier in determinate cultivars, which do not continue to produce new foliage throughout the season. Determinate cultivars complete their growth cycle in 80 to 100 days, even if nutrient availability and environment continue to be favorable. These cultivars may exhibit indeterminate-type growth if grown under warm night conditions, when minimum temperatures do not fall below 60° to 65°F (15.6° to 18.3°C).

Stage III: Tuber Initiation

The onset of tuber initiation (tuberization) is controlled by growth-regulating hormones produced in the potato plant.

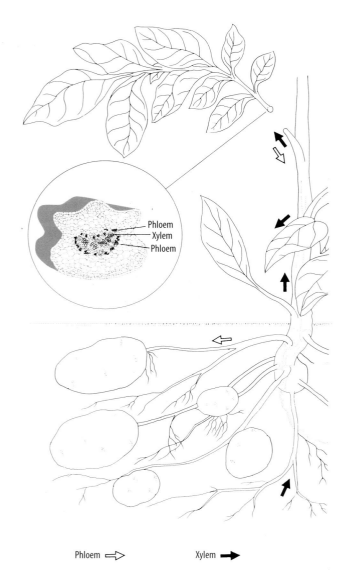

Phloem
Xylem
Phloem

Phloem ⇨   Xylem ➡

Figure 5. Diagram of the conducting tissues (vascular system) of a potato plant. The xylem carries water and nutrients from the roots, and the phloem distributes the products of photosynthesis.

Before tuber initiation can start, more carbohydrate must be produced by photosynthesis than is needed to support the growth of leaves, stems, and roots. The excess carbohydrate, in the form of sucrose, is translocated down to stolon tips, where tuber initiation begins. Tuber initiation usually begins when the leaf area index (the ratio of total plant leaf area to the area of soil surface covered by the plant) is 1.5 to 2.0. In indeterminate cultivars such as Russet Burbank this occurs at the 8- to 12-leaf stage; determinate cultivars such as Russet Norkotah may begin to initiate tubers earlier. Tuber initiation is affected by soil moisture, soil temperature, and nitrogen management.

Tubers usually first appear on the lower, older stolons, when high amounts of sucrose begin to build up in the stolon tips. During tuber initiation the area just behind the stolon tip undergoes cell division, and many of the axillary buds become eyes. Sucrose can be converted into starch for storage in the expanding tuber cells before tuber initiation is visible. Under normal conditions, Russet Burbank tubers that will subsequently reach harvestable size are initiated over a 10- to 14-day period, and the tuber initiation phase is completed by about the 12-leaf stage. Indeterminate cultivars may continue to initiate tubers after the end of this growth phase, but these late tubers may not reach harvestable size.

Stage IV: Tuber Growth

Although some cell division occurs, tuber growth is primarily a period when cells formed during the initiation phase expand 7- to 18-fold with water and starch reserves. Tubers become the dominant sink for the carbohydrates produced by photosynthesis and for inorganic nutrients such as nitrogen, phosphorus, and potassium. Carbohydrates move into the expanding tubers primarily as sucrose, which is rapidly converted into starch. The microscopic starch grains formed make up 10% to 25% of the fresh weight of a tuber. As tuber cells accumulate starch, the ratio of dry matter content to water content increases. Tuber dry matter content is measured as specific gravity; the higher the dry matter, the higher the specific gravity. Higher specific gravity is more desirable, especially in tubers to be used for processing.

If disease, inadequate fertilization, or other factors reduce the ability of the plant to take up inorganic nutrients, nutrients are drawn from the vines to support continued tuber growth. The vines then begin to show symptoms of aging (senescence) prematurely. They become more susceptible to diseases such as Verticillium wilt, early blight, and black dot.

Because tubers are modified stems, they have the same kinds of structures as aboveground stems (Fig. 6). Each eye of a tuber contains several axillary buds that can give rise to new stems. Normally one bud is dominant and is the only one that will sprout. However, if this sprout is removed or severely damaged, other buds within the eye will sprout.

The periderm (skin) of a tuber is made up of a single layer of corky cells called the epidermis and several layers of sloughed-off corky cells above the epidermis. When tubers are injured, by bruising or cutting for example, a wound periderm forms. It consists of several layers of corky cells containing a high proportion of a waxy substance called suberin; the formation of wound periderm is called suberization. Suberin contains compounds toxic to microorganisms and acts as a barrier to invasion by disease organisms. Suberization is important in protecting stored tubers and seed pieces from infection.

Lenticels are surface pores present on both stems and tubers (see Fig. 6). They are the same kind of structure as leaf stomata, and, like stomata, they serve as sites for gas exchange. Layers of suberin are present underneath the lenticels and normally act to prevent infection. However, under conditions of high moisture, cells in the underlying cortex layer may swell and break through the suberin layers, causing open or enlarged lenticels. Enlarged lenticels are susceptible to disease infection.

The tuber has a vascular system, part of which is concentrated in an area called the vascular ring and part of which is dispersed throughout the tuber. The tuber xylem carries water and the phloem carries carbohydrates and other nutrients to expanding cells. Certain physiological conditions or disease organisms can cause portions of the tuber vascular system to turn dark brown. Tuber tissue lying between the vascular ring and the periderm is called the cortex; tissue inside the vascular ring is called the medulla. A watery region called the pith is present in the center of the medulla, with rays extending outward to the eyes.

Stage V: Maturation

During the maturation period, tuber skin (periderm) thickens, or sets, the tuber dry matter content reaches a maximum, and the vines senesce. Senescing vines begin to turn yellow and lose leaves, and photosynthesis decreases. Increases in tuber dry weight during this period are mainly due to movement of carbohydrates from vines into tubers. During this time eyes become dormant, and starch synthesis decreases and then finally stops when vines die. Untimely late-season release or application of nitrogen may stimulate indeterminate cultivars to form additional foliage, which requires energy for growth. This demand may cause a loss of starches already deposited in tubers, reducing tuber dry matter. The periderm continues to thicken after vine death.

Energy Resources and Yield

Producing a high yield of quality tubers at reasonable cost requires the right balance between the supply of energy produced by the plant and the demand placed on this energy supply. The energy supply for a potato plant consists of

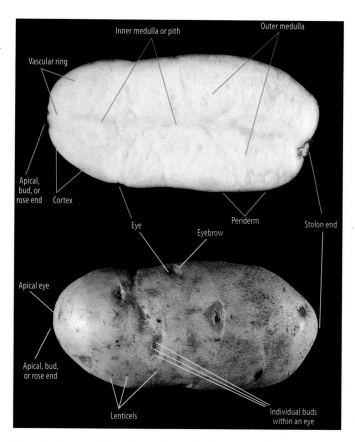

Figure 6. Longitudinal section and exterior of a potato tuber, showing structural features. Xylem and phloem are located in the vascular ring and throughout the medulla tissue. After the dormancy period, eyes gain the ability to form sprouts, starting at the apical eye and proceeding from the bud to the stolon end.

carbohydrates produced by photosynthesis. Carbohydrates supply energy needed for two basic functions: respiration and growth.

Respiration, the process that supplies energy needed to support the metabolism of living cells, always has first priority in its demand for energy. The energy supply remaining after respiration needs are met is available for tissue expansion and deposition as reserves. However, plants cannot support the growth of roots, stems, leaves, and tubers at uniform rates simultaneously. Thus, the energy supply is allocated according to a system of priorities (Fig. 7).

During the vegetative phase, growth of roots, stolons, stems, and leaves has priority for energy after respiration. The plant builds the vegetative framework needed to synthesize and distribute carbohydrates necessary for tuber growth. Once its leaf area index reaches 3 to 4, a potato plant is able to produce all the energy it needs to sustain maximum tuber growth.

When the foliage accumulates more carbohydrate than is needed to support vine growth, the surplus is available for tuber initiation and growth. As more tubers are initiated and

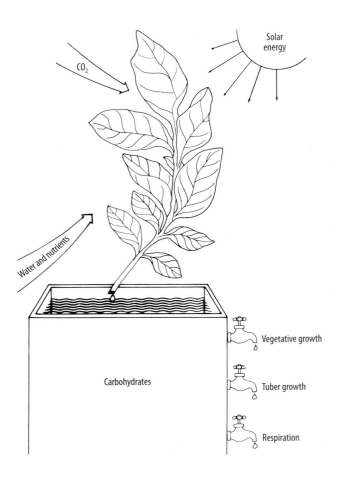

Figure 7. Carbohydrates produced in the leaves by photosynthesis are an energy supply that supports plant functions. As demand drains the pool of energy, vegetative growth ceases first, then tuber growth, and finally respiration.

begin to expand, tuber growth takes precedence over vegetative growth and initiation of new tubers. By far the greatest amount of tissue production occurs in tubers (Fig. 8). The rate of tuber growth depends on the availability of carbohydrate in excess of that required for foliar growth and plant metabolism. The more carbohydrate produced by the vines, the faster tubers grow; maximum tuber growth is supported as long as the leaf area index is above 3.0 and conditions are favorable.

If excess nitrogen is applied to indeterminate cultivars at planting, excess foliar growth occurs and the onset of tuber growth may be delayed by as many as 10 to 14 days. Although maximum leaf area index increases, tuber growth rate does not increase. Therefore, the delayed tuber growth may result in reduced yields if the season is too short to allow completion of the growth cycle. Excess nitrogen has less of an effect on determinate cultivars.

Effects of Pest Injury on Yield

Pest injury often affects the production, transport, or allocation of energy resources within the potato plant. When pests interfere with these important plant functions, they may severely limit yields. Weeds may reduce yields by competing with potato plants for available water and nutrients.

Water and Nutrient Transport. Nutrients and water move from plant roots to the leaves and stems through xylem tissue within the vascular system. Certain pathogens restrict flow within the xylem, interfering with water and nutrient transport and causing wilting and premature aging of potato vines. Verticillium wilt and blackleg are the most important diseases causing xylem blockage. Bacterial ring rot and severe Rhizoctonia infections are also very effective in reducing xylem transport.

Carbohydrate Transport. A number of diseases may reduce or block the transport of carbohydrates from foliage to tubers by clogging the phloem tissues of the vascular system. Small green tubers called aerial tubers form at the base of stems or in axillary buds when sugar accumulates because it cannot be transported into tubers fast enough. Sucrose also accumulates in leaves, causing them to thicken and become brittle and papery. Leaves and stems of red-skinned varieties turn reddish. Among the pathogens that can cause these symptoms are *Potato leafroll virus* (aerial tubers are rarely caused by leafroll virus), blackleg, Rhizoctonia, *Beet curly top virus*, and the phytoplasmas that cause aster yellows and witches' broom. Conditions that reduce movement within the phloem, such as waterlogging of the roots, can also cause similar symptoms.

Foliage Reduction. Some pests affect crop production by reducing the leaf area available to intercept sunlight and perform photosynthesis. Because fully developed potato vines have more leaves than required for maximum tuber growth, vines can sustain some loss in leaf area before a reduction in photosynthesis limits tuber growth. However, foliage losses that do not reduce photosynthetic output may cause the plant to accelerate vine growth at the expense of tuber growth, delaying or reducing yields.

The stage of growth when foliar damage occurs is critical. If foliage loss occurs before adequate leaf area is achieved, it may be difficult for plants to recover; tuber production may be delayed or prevented. Damage that occurs during tuber initiation or growth causes a greater loss in yield than damage occurring during the maturation phase. For example, early blight often occurs late in the season, when the loss of foliage for photosynthesis does not result in significant yield reduction. However, if early blight develops soon enough, controls are needed to prevent yield loss. Foliage feeding by Colorado potato beetles is also much more serious when it occurs early in the season.

Second Growth

If water uptake is less than water loss from transpiration during hot weather, water stress in leaves causes stomata (leaf pores) to close. Photosynthesis then declines as carbon dioxide availability is reduced by stomatal closure. This interrupts the flow of carbohydrate that maintains normal tuber growth. When conditions again become favorable for photosynthesis and carbohydrate levels increase, improving the supply to tubers, growth of individual tubers may resume unevenly. Some areas of a tuber may grow more slowly than others; some areas may not resume growth. Uneven tuber growth results in various types of tuber malformation, called second growth disorders. The parts of tubers that grow slowly or do not resume growth may have lower starch concentrations and higher sugar levels. Such tubers are unacceptable for processing. Russet Burbank is highly susceptible to second growth disorders; most other cultivars grown in the western states are fairly resistant.

Storage

Tubers continue to respire after harvest. Energy derived from respiration is required to support suberization during the curing period and to support metabolic processes that continue in storage. Adequate and uniform storage ventilation is necessary to provide oxygen for respiration, to control humidity, and to remove the heat, water, and carbon dioxide produced by tuber respiration.

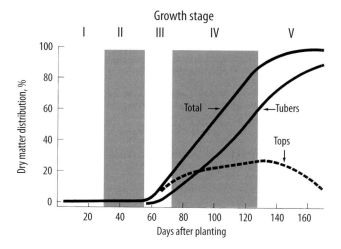

Figure 8. Distribution of dry matter production between tops (vines) and tubers of Russet Burbank potato plants. These curves are characteristic of indeterminate cultivars. Determinate cultivars such as Russet Norkotah have similar curves, but the growth stages begin earlier and are completed more rapidly. The actual timing of growth cycle events depends on local growing conditions.

Sugar Accumulation

When tubers are stored at temperatures lower than 45° to 48°F (7.2° to 8.9°C), depending on the cultivar, some starch is converted into sugars ("cold sweetening"). The lower the temperature, the faster this conversion occurs. Sugar accumulation is undesirable in tubers used for processing because the sugars caramelize during processing, darkening the product. For this reason, tubers to be used for chips usually are stored at no less than 50°F (10.0°C), and tubers used for french fries no less than 45°F (7.2°C). If sugars do accumulate in storage, their levels may be reduced by holding tubers at temperatures above 50°F (10.0°C) for 1 or 2 weeks before processing, a procedure called reconditioning. Some varieties are less likely to accumulate sugars or are more easily reconditioned, making them more desirable for processing. An increasing number of cultivars are resistant to cold sweetening. These can be stored at 42° to 44°F (5.5° to 6.7°C) if to be used for French fries; at 42° to 45°F (5.5° to 7.2°C) if to be used for chipping.

Dormancy

After potato tubers are harvested, a certain length of time is required before the eyes can sprout. This time is called the dormant period; its length depends on cultivar and conditions during growth and storage. Hot weather during growth, nitrogen deficiency, and high or fluctuating storage temperatures shorten the dormant period. At the end of the dormant period the apical eye at the bud (rose) end of the tuber of most cultivars is dominant over the others; it will sprout first, followed by the remaining buds. In some cultivars, many eyes sprout simultaneously. This apical dominance weakens with time after dormancy is over, until eventually all eyes will sprout simultaneously. Seed tubers are generally stored at 35° to 38°F (1.7° to 3.3°C) to prolong dormancy and delay sprouting.

Aging

The respiration that takes place during storage slowly uses up some of the starch reserves that are needed to support development of sprouts from planted seed tubers. Other changes take place as well during storage, including the loss of dormancy and apical dominance and a reduction in cell membrane function and integrity. The net effect of these changes is an aging of tubers that is sometimes called physiological aging. Some aging of seed tubers is desirable to achieve proper sprouting of seed pieces. Physiologically older seed tubers produce more stems per hill and set more tubers that tend to be smaller. However, if seed tubers are excessively aged, weak plants may be produced or "little tuber" disorders may result. Little tuber, also called "no top," occurs when stolons emerge directly from eyes and produce one to several small tubers but no aboveground stems. The small tubers produced often are called "submarine tubers."

Managing Pests in Potatoes

No single management program is suitable for all potato crops. Pest problems vary from field to field and season to season because of differences in soil type, cropping history, cultural practices, cultivars, and the nature of surrounding land. Choice of market and market conditions also affect the feasibility of management options because they determine how a crop must be handled and the value of that crop. Regardless of conditions, however, four components are essential to any IPM program:

- accurate identification of pests and beneficials
- field monitoring
- control action guidelines
- effective management methods

The purpose of this chapter is to help you incorporate these components in a program tailored to the specific needs of each field. Management practices that are important in controlling potato pests are summarized in Table 3.

Pest Identification

Because most pest management options, including pesticides, are effective only against certain pest species, you must know how to identify the pests that are present or are likely to appear. Different control methods may be needed even for closely related species.

Use the descriptions and photos in this manual to identify pests commonly affecting potatoes in the western United States. Check the suggested reading for other sources of information. Remember that the help of experienced professionals is required to reliably diagnose some pest problems; DO NOT HESITATE TO SEEK TECHNICAL HELP IF YOU ARE NOT CONFIDENT OF THE ACCURACY OF AN IDENTIFICATION. Extension agents, farm advisors, or other experts in your area can help you identify pests and can direct you to other specialists if necessary.

Monitoring

By monitoring your field, you can get the information you need to make management decisions. Monitoring includes keeping records of weather, crop development, and management practices, as well as evaluating incidence and levels of pest infestations. Seasonal monitoring activities are summarized in Figure 9.

ACTIVITY	Pages in text	Previous crop	Between crops	Preplant	Planting to emergence	Vegetative growth	Tuber initiation	Tuber growth	Maturation and harvest	Storage
Aerial monitoring	16									
Soil and water analysis	15, 28, 30									
Soil sampling for nematodes	119									
Weed surveys	126									
Monitor for wireworms	62									
Keep weather records	13, 88									
Petiole nutrient analysis	30									
Keep water budget	28									
Monitor soil moisture	27, 32									
Monitor soil temperature	32, 81, 122									
Monitor weed emergence	126, 130									
Monitor for Colorado potato beetle	55									
Monitor for aphids	48				1					
Monitor for psyllids	52									
Monitor for loopers	57									
Monitor for tuberworm	60									
Monitor for early blight and late blight	86, 89									
Monitor for white mold	91									
Monitor canopy moisture, temperature, and humidity	13, 86, 90									
Monitor temperature and humidity	33, 74, 80, 102									

1. In warmer areas where aphids are active early in the season or all year.

Figure 9. Monitoring activities important for potato pest management.

Weather Data

Because weather influences the development of the potato plant and its pests, reliable weather data is critical to many crop and pest management decisions. For example, control actions for some disease problems must be made using forecasting models before symptoms appear in the field. Models that predict outbreaks of late blight rely on measurements of rainfall, temperature, and in some cases relative humidity and leaf wetness. Daily high and low temperatures are needed if degree-day accumulations are used to schedule early blight control. Evapotranspiration data may be useful for scheduling irrigations, and weather forecasts are essential for scheduling planting, harvest, frost protection irrigation, chemical applications, and other operations.

Many newspapers and radio stations in agricultural areas report such information; the National Weather Service broadcasts weather information on NOAA Weather Radio, VHF channels at 162.42, 162.50, and 162.55 MHz. Evapotranspiration information for California locations is available from the California Department of Water Resources' CIMIS program. Daily weather information for a number of locations throughout California is available from the University of California Statewide IPM Program (UC IPM). Evapotranspiration and other weather information is available for locations throughout the Pacific Northwest and some locations in Montana, Nevada, and Wyoming from the Pacific Northwest Cooperative Agricultural Weather Network (AgriMet). See the suggested reading for ways to access CIMIS, UC IPM, and AgriMet, all of which are available via the World Wide Web.

Regional weather information may not reflect the situation in your field, which is affected by local variations in

Figure 10. A weather station placed in or near your fields will give you the most accurate weather information for making management decisions.

Table 3. Summary of Management Activities Important for Controlling Potato Pests.

PEST COMPLEX	MANAGEMENT ACTIVITIES	PEST COMPLEX	MANAGEMENT ACTIVITIES
PREVIOUS CROP		**PREPLANT**	
Seed tubers	Visit seed production areas to observe seed fields.	Insects and related pests	Control weed hosts.
	Mature seed crop with use of vine-killing agents or vine removal if necessary.		Use soil-applied or seed-treatment insecticides when necessary.
	Harvest seed tubers under optimum temperature and soil conditions, with properly cleaned equipment, and with minimum mechanical damage.	Diseases	Use highest-quality certified seed tubers.
			Warm seed potatoes to proper temperature before cutting.
	Harvest seed tubers before exposure to freezing temperatures and before soil temperature falls below 50° F (10.0° C).		Suberize cut seed or plant immediately after cutting.
			Use seed treatment fungicides when necessary.
	Store seed tubers carefully to maximize their vigor and health.		Provide proper planting conditions:
Insects and related pests	Destroy weed host sources.		• Correct soil structure
	Use rotations that suppress potato insects that survive in the soil such as Colorado potato beetle and wireworms.		• Soil temperatures above 45°F (7.2°C)
			Establish soil moisture that will remain adequate until emergence.
Diseases	Destroy weed hosts.	Physiological disorders	Store seed tubers and cut seed properly.
	Destroy potato volunteers.		Obtain desirable seed piece size.
	Use rotations that help reduce inoculum of soilborne diseases that may be present such as Verticillium wilt, Rhizoctonia, bacterial ring rot, silver scurf, and black dot.		Provide proper planting conditions.
		AT PLANTING	
Nematodes	Use rotations that suppress the root knot nematode species and other harmful species that are present.	Insects and related pests	Use systemic insecticides properly when necessary.
		Diseases	Use good sanitation during seed cutting and planting.
Weeds	Use rotations that allow control of problem weeds.		Sample and monitor for diseased seed pieces.
	Avoid herbicides that leave residues harmful to potatoes the following year.		Plant at correct depth in warm (above 45°F [7.2°C]), moist soil.
BETWEEN CROPS		Physiological disorders	Provide proper soil fertility.
Seed tubers	Store seed tubers in proper environment.		Plant at correct depth and spacing.
	Obtain North American Certified Seed Potato Health Certificate for prospective seed lots.	**PREEMERGENCE**	
		Diseases	Avoid preemergence irrigations.
	Purchase seed lots with winter grow-out test results.		Plant shallowly to hasten emergence; form hills after emergence.
Insects and related pests	Destroy weed host sources.	Weeds	Apply preemergence herbicides when necessary.
	Monitor for wireworms.	**EMERGENCE THROUGH TUBER GROWTH**	
Diseases	Eliminate cull piles.	Insects and related pests	Monitor for insects and mites.
	Destroy potato volunteers.		Hill to prevent exposure of tubers to pests.
	Assay soil for soilborne pathogens.		Apply soil or foliar pesticides when necessary.
	Use soil fumigation when necessary and feasible.	Diseases	Maintain proper soil moisture and fertility.
Physiological disorders	Analyze for soil fertility, chemistry, and physical constraints.		Avoid extended periods of wet foliage.
			Monitor for disease development.
Nematodes	Use tillage to destroy crop residues.		Apply fungicides when necessary.
	Assay soil for nematode species.	Physiological disorders	Maintain proper soil moisture and fertility.
	Use fumigation when necessary and feasible.		Hill to prevent exposure of tubers to heat and light.
Weeds	Use tillage and herbicides to control perennials and potato volunteers.	Nematodes	Assay soil for nematodes and apply nematicides if necessary.
	Allow breakdown of harmful herbicide residues.	Weeds	Monitor for weed emergence.
			Time hilling to control emerging weeds.
			Apply postemergence herbicides when necessary.
		Sprout inhibition	Apply sprout inhibitors to healthy plants as needed.

PEST COMPLEX	MANAGEMENT ACTIVITIES
MATURATION AND HARVEST	
Insects and related pests	Keep tubers covered to avoid tuberworm damage and greening.
	Use early and rapid vinekill to avoid aphid migration and frost.
Diseases	Maintain proper soil moisture.
	Apply foliar fungicides or vine-killing agents when necessary for control of late blight.
	Harvest after complete vine death for tubers destined for storage and to reduce late blight tuber infections.
	Use careful, bruise-reducing harvesting and handling procedures.
	Harvest at proper soil moisture and temperature.
	Maximize soil and organic matter removal before storage.
	Avoid repeated pile disturbance and fragmented unloading of bays to reduce chances of spreading silver scurf.
Nematodes	Assay soil for nematodes and apply nematicides if necessary.
	Time harvest to avoid tuber damage when possible.
	Isolate lots with badly infested tubers.
Weeds	Apply contact herbicides for complete vinekill and control of nutsedge.
	Manage water for late-season weed control.
STORAGE	
Diseases	Use proper sanitation in storage areas.
	Use proper curing conditions.
	Use ventilation and reduced humidification to hasten drying of wet tubers.
	Maintain proper ventilation, humidity, and temperature.
	Monitor storages for disease development.
	Avoid repeated pile disturbance and fragmented unloading of bays to reduce chances of spreading silver scurf.
Physiological disorders	Maintain proper ventilation, humidity, and temperature.
Nematodes	Monitor tubers going into and coming out of storage for symptoms of nematode damage.
	Store at temperatures that prevent nematode multiplication if compatible with intended use.

terrain, vegetation, elevation, and other conditions. It is best to collect your own weather data with instruments placed in or near your fields (Fig. 10). Weather instruments can range from simple devices such as a maximum-minimum thermometer to electronic devices that continuously monitor and record weather information for transfer to a computer. Even more sophisticated stations automatically transmit the data to a remote computer. Some relatively simple devices are available that keep track of temperatures and accumulated degree-days. Set up and maintain the weather instruments according to manufacturers' instructions and calibrate them regularly to ensure accuracy. Keep records of all your observations.

Pest Activity

Monitoring during the growing season is devoted mostly to checking for insect populations and appearance of diseases. This information may be used to make treatment decisions or to follow up on pest management practices. Certain problem weeds such as dodder may require frequent monitoring in specific areas. Visit each field at least once a week; during critical periods of crop development or when a population is approaching a treatment threshold, visit the field every 2 or 3 days. In each growing area, the emphasis is usually on a few pest species that occur regularly and threaten yield or quality. Work your sampling methods into a single routine so that you do not have to make a special pass through the field for each one. In seed potato fields keep foot traffic to a minimum to reduce the spread of diseases that are mechanically transmitted. Always disinfect footgear before entering seed potato fields.

Other pest monitoring activities are needed only once or a few times each season. For instance, conduct a weed survey late in the season of the previous crop, and repeat the survey once or twice during the growing season and again at harvest. In areas where early dying or nematodes are a problem, collect soil samples for laboratory analysis or extraction of nematodes and potential pathogens the fall before potatoes are planted to help plan control measures. Samples should be taken before the soil profile has been allowed to dry out.

Soil, Water, and Tissue Testing

Contact a reliable laboratory to test soil, water, and plant tissue. Preplant soil analysis helps determine fertilizer requirements. Soil tests can also measure salinity, soil reaction (pH), organic matter content, and other factors that affect crop growth and management practices.

Occasional testing of irrigation water helps detect harmful increases in salinity, especially where the water comes from wells or where surface water is known to be high in salts. Tests should include sodium, zinc, and boron levels, and also the concentration of total salts. References listed under "Soil and Water" in the suggested reading contain information on how to interpret test results and deal with salinity problems. Tests should also measure the level of nitrates in the irrigation water, information that is valuable for planning your nitrogen fertility program, particularly in the western states where the nitrate level in many wells is increasing.

Analysis of plant tissue is the best way to monitor the crop nutrient status during the growing season. Take at least three or four petiole samples and associated soil samples during the period from vegetative growth (growth stage II) through tuber maturation (growth stage V). These samples give you considerable information about the health of the crop and fertility deficiencies, excesses, or imbalances. The results are useful for planning the next season's preplant fertilizer applications, and early analyses can assist in current-season nitrogen management.

Aerial Monitoring

Aerial infrared (IR) photography may be useful for soil mapping and quickly assessing the status of general plant health over large areas. It can display areas in the field stressed by factors such as inadequate moisture, poor soil fertility, insects, or diseases, and help you identify sites where ground monitoring is necessary. If this service is available in your area, check with your Extension Service and with the operator who provides the service to see what this type of monitoring may do for you. Make arrangements to have your fields photographed early and have analyses of the photographs provided regularly.

Keeping Records

Knowing what happened in a field last year and in the last potato crop that was grown there, as well as what happened in adjacent fields, can provide valuable information for making current-season management decisions. Keep a file, notebook, or computer record of the following for each field:

- weekly monitoring reports
- weed surveys
- records of fertilizer and pesticide applications, including names of materials, rates, dates, and methods of application
- laboratory reports
- aerial photographs with reports attached
- agronomic information, including crops and cultivars planted, planting and harvest dates, yields, and grade out
- disease reports
- problem areas

In addition to maintaining records that apply to specific fields, keep a file of local weather data, including charts of accumulated degree-days if you are using them to make pest control decisions.

Control Action Guidelines

Control action guidelines indicate when management actions are needed to avoid losses due to pests or other stresses. Guidelines for some insect pests are numerical thresholds based on specific sampling techniques; they are intended to reflect the population level that will cause economic damage if left unchecked. Guidelines for other pests, including most pathogens and weeds, are usually based on the history of a field or region, cultivar being grown, the stage of crop development, observed symptoms or damage, weather conditions, and other observations.

These guidelines often change as new cultivars and cultural practices are introduced and as new information on pests becomes available. The guidelines presented in this manual will be revised as more research is completed on potato pest management.

Management Methods

The ideal IPM program protects the crop while interfering as little as possible with long-term maintenance of the production system. The least costly, most reliable way to deal with pest problems is to anticipate and avoid them. When pesticides are needed, choose materials and application methods that control pests effectively with minimum impact on beneficials, nontarget organisms, and the environment.

Seed Quality and Seed Certification

Many pests can be transmitted in infected seed tubers, including potato viruses, bacterial ring rot, blackleg, late blight, Rhizoctonia, common scab, silver scurf, wilt diseases, and root knot nematodes. Stem cutting and micropropagation techniques have been developed to obtain pest-free potato plants for propagation and production of certified seed tubers. Disease-free stem cuttings or tiny pieces of meristem tissue are cultured and propagated under sterile conditions to produce large numbers of disease-free plantlets or minitubers (Figs. 11 and 12). The majority of disease-free plantlets are

Figure 11. In the first stages of certified seed potato production, tissue culture techniques are used to produce disease-free potato plantlets and microtubers.

LARRY L. STRAND

Figure 12. Small tubers produced on plants grown under greenhouse conditions may be used to grow the first generation of seed potatoes in the field.

produced from meristem tissue cultures in public or private laboratories. Several generations of plants are grown in the field to produce certified seed tubers that will be sold to commercial growers.

Certified seed tubers are not guaranteed to be disease free. They are certified to have shown no more than certain low percentages of pest and disorder symptoms during the inspections required by a state's seed certification program. The allowable level of symptom expression for each pest or disorder is called a tolerance level, and these levels vary from state to state. A zero tolerance exists for certain pests, such as bacterial ring rot and root knot nematode. To pass these tolerances, seed lots must be inspected at least twice in the field during the growing season and be inspected in storage or at the time of shipment. Samples of each seed lot are grown and inspected in winter field trials in warm locations such as southern California and Hawaii, or in winter greenhouse plantings. Pests for which tolerances are enforced and the range of tolerances among the western states are listed in Table 4. More detailed information about seed certification programs in individual states can be found in references listed in the suggested reading.

Extra precautions are taken to reduce the incidence and spread of pests in fields where seed potatoes are grown. Most seed potatoes are grown in cool, short-season areas (see Fig. 2 in the first chapter) where long rotations are used between potato crops and where pest populations, including vectors of potato viruses, remain low. Symptoms of virus diseases and other diseases are visible in these areas, and seed growers routinely rogue seed lots to remove diseased plants. Ring rot symptoms also are visible under these conditions and can be detected by trained certification inspectors. Ring rot is a

zero tolerance disease—roguing is not allowed, and seed lots confirmed to have ring rot are rejected from certification. Ideally, seed fields are isolated from commercial fields, wild plums, stone fruit orchards, and home gardens, which may harbor aphids that transmit potato viruses.

Clean and sanitize all equipment thoroughly before entering seed fields and keep mechanical contact to a minimum to reduce spread of certain pathogens. Minimize cultivation, avoid cultivation when plants are tall enough to be brushed by equipment, use herbicides instead of cultivation to control weeds, use solid set or self-moving sprinklers if possible, and keep people and animals out of fields as much as possible. Whenever work in a field is required, do it when the vines are dry; mechanical transmission of pathogens occurs more readily when vines are wet. Remove diseased plants, including any tubers that are present, as early in the season as possible. Diseased plants, and adjacent plants that are in contact with them, should be removed when they are dry and immediately placed in plastic or other smooth, nonabrasive bags to keep them from touching other plants. Fields with a history of ring rot, root knot nematode, or certain other problems may not be used for certified seed tuber production in some states until nonhost crops have been grown for a specified number of years. Sanitize shoes and equipment between use in different fields or with different seed lots.

Aphid control programs are designed not only to prevent the development of aphid populations in seed fields, but also to prevent populations from building up elsewhere and moving into seed fields. Communitywide programs may be

Table 4. Range of Tolerance Levels for Latest Generation of Seed in Western States Seed Certification Program.

Disease symptom	Range of tolerance level (%)
leafroll	0.2–0.4
PVX	1.0–4.0
mosaic virus	0.5–2.0
calico (AMV)	0.5–1.0
Total all viruses	1.0–3.0 [a]
haywire	0.5–2.0
witches' broom	0.5–1.0
spindle tuber	0.0–0.1
blackleg	0.25–0.4
bacterial ring rot	0.0
late blight	1.0 [b]
giant hill	0.5 [c]
root knot nematode	0.0
varietal mix	0.1–1.5

a. Does not include PVX
b. Shipping point tolerance
c. Colorado, Montana

necessary in some seed-growing areas to prevent buildup of aphid vectors in commercial fields or gardens, or on alternate hosts such as stone fruits. In some areas systemic insecticides are applied at planting or, where seasons are longer, when about 75% of the plants have emerged. Insecticide sprays are used if aphids appear in seed fields. Controlling mustards and other weed hosts of aphid vectors in field borders and adjacent areas helps prevent the development of potentially damaging aphid populations. Planting border crops around small seed potato fields (less than 1 acre, or 0.4 ha) can reduce the spread of styletborne viruses such as PVY. These viruses are lost when infected aphids probe the border crop before moving into the potato plants. The border crop can be any plant that is not a host for potato viruses and that will stay green throughout the potato growing season. Fallow ground on all sides of the crop border increases its attractiveness to aphids.

Be familiar with the certification requirements that apply to the seed tubers you buy. Each state uses colored tags to identify the quality of seed tuber lots, and printing on the tags identifies generation; learn what these tags mean. Obtain seed tubers from a source with a reputation for supplying good-quality seed. If possible, visit the areas where the seed is grown to investigate growing conditions, cultural practices, and overall field condition, all of which influence seed quality. Request a North American Certified Seed Potato Health Certificate for seed lots you are considering. The Health Certificate reports the results of winter and summer test readings for a seed lot and contains the pedigree of the seed lot. The certificate also contains farm history information, which lists whether ring rot or late blight have been found on the farm within the last 10 years. A Health Certificate can be obtained from the grower or state certification agency. Links to a copy of the North American Certified Seed Potato Health Certificate can be found on the web sites for the Potato Association of America (www.umaine. edu/paa) and Potato Information Exchange (oregonstate. edu/potatoes).

Follow careful handling and sanitation procedures to keep certified seed tubers from becoming contaminated. Holding seed tubers under proper conditions for suberization and keeping seed-cutting equipment clean and sanitary are worthwhile precautions to protect your investment in good-quality seed tubers, as well as the time and expense invested in growing the crop.

Biological Control

Any activity of a parasite, predator, or pathogen that keeps a pest population lower than it would be otherwise is considered biological control. One of the first assessments that should be made in an IPM program is the potential role of natural enemies in controlling pests. Control by natural enemies can be inexpensive (when populations are native), effective, self perpetuating, and not disruptive of natural balances in the crop ecosystem.

The use of broad-spectrum insecticides in potatoes reduces the numbers and hence the effectiveness of many natural enemies in potato fields. Because of certification requirements and the potential for virus spread into potato fields, damage thresholds in seed potato fields and in commercial fields in many areas are so low that the activity of natural enemies would not keep insect populations sufficiently controlled. However, natural predators in field borders or adjacent wild areas where insecticides are not used may serve to keep reservoirs of certain pests such as aphids and Colorado potato beetle at lower levels.

To gain the greatest benefit from biological control of insect pests, choose insecticides, rates, and application methods that have a minimal impact on pests' natural enemies. Growing certain green manure cover crops in rotations and incorporating their residue reduces soil populations of harmful nematode species partly by encouraging the buildup of fungi and other natural enemies that attack nematodes.

Resistant Cultivars

Plant breeding is one of the most powerful tools available for both the management of pests and the production of high-quality yields. Pest management is one of many factors that must be taken into account when choosing cultivars; characteristics of cultivars grown in the West are listed in Table 5.

Cultivars tolerant or resistant to disease can provide long-term, economical protection from pathogens that otherwise could inflict severe losses every season. A plant is considered resistant to a disease if the disease organism cannot infect the plant, or if, on infection, the organism does not cause any disease symptom or measurable effect on yield or quality. Pest resistance is a matter of degree, ranging from highly resistant to highly susceptible (Table 6). Cultivars often are ranked relative to one another—one cultivar being more resistant or more susceptible to a given disease than another. The term "tolerant" sometimes indicates a degree of resistance that allows the plant to tolerate a certain amount of disease with no more than moderate losses. For example, the cultivar Ranger Russet may be considered tolerant to Verticillium wilt; unless the disease is severe, infected plants produce satisfactory yields as long as optimum growing conditions are provided.

Part of every breeding program is the search for resistance to serious diseases, disorders, and nematode pests. Resistance to insect pests is being investigated. New potato breeding selections are assessed for resistance to several viruses, leafroll net necrosis, root knot nematodes, Verticillium wilt, common scab, blackleg, early blight, late blight, and several physiological disorders.

The potato cultivars commonly grown in the western United States differ in their relative resistance to important

Table 5. Characteristics of Potato Cultivars in the Western United States.

Cultivar	Maturity	Dormancy[1]	Tuber	Principal uses	Possible problems
Alturas	very late	3	oval to oblong; light russet; shallow eyes	processing	very late tuber bulking; necrotic strains of PVY
Atlantic	medium	3	oval; scaly skin	chipping	internal brown spot; hollow heart; metribuzin injury
Bannock Russet	late	4	oblong; dark russet skin	processing, fresh market	late tuber bulking; shatter bruise; large tuber size
CalWhite	medium	2	oblong; white skin	processing, fresh market	heat sprouting; common scab; leafroll net necrosis; metribuzin injury
Cherry Red	early to medium	3	oval to oblong; red skin; shallow eyes	fresh market	metribuzin injury; large tubers
Chieftain	medium	2–3	oval; medium red skin	fresh market	light skin color; deep eyes; large tuber size
Chipeta	late	4	round; patchy skin	chipping	Rhizoctonia injury; late tuber bulking; large tuber size
Defender	late	3	long; white	processing	common scab
Gem Russet	late	4	long; russet skin	processing, fresh market	latent PVY expression; blackspot bruise; slow emergence
Norland	very early	2	round; red skin; shallow eyes	fresh market, home garden	color loss and sprouting in storage; pinto coloring
Ranger Russet	late	3	long; russet	processing, fresh market	blackspot bruise; submarine tubers on aged seed
Red LaSoda	early to medium	3	oval; light red skin; deep eyes	fresh market, home garden; heat tolerant	internal necrosis; hollow heart; deep eyes
Russet Burbank	late	4	long; russet; shallow eyes	fresh market, processing; stores well	malformed tubers; sugar end; leafroll net necrosis
Russet Norkotah	early	3	long; russet; shallow eyes	fresh market	early dying; latent PVY expression
Russet Norkotah strains	early to medium	3	long; russet; shallow eyes	fresh market	early dying; latent PVY expression
Russet Nugget	late	3	oblong; russet; shallow eyes	processing, fresh market	elephant hide in saline soils when nitrogen and water excessive
Sangre	medium	3–4	oblong; shallow eyes; dark red skin	fresh market; stores well	slow emergence; skin russeting
Shepody	medium early	3	long-oblong; white skin; shallow eyes	processing	metribuzin injury; large tubers; latent PVY expression
Umatilla Russet	late	3–4	long; russet; shallow eyes	processing, fresh market	Fusarium dry rot; shatter bruise; pressure bruise
White Rose	early to medium	2–3	long; white skin; medium eyes	fresh market; heat tolerant	common scab; thin skin; short shelf life; greening
Yukon Gold	medium	2	round; white skin; yellow flesh	fresh market	seed and storage rot; low yields; early dying

1. Months required at 40° F (4.4° C) to break dormancy of seed tubers grown under normal conditions.

diseases and disorders (see Table 6). Potato selections with high resistance to physiological disorders, net necrosis, PVY, PVX, late blight, Verticillium wilt, or root knot nematodes are now available.

In recent years, recombinant DNA technology has been used to create transgenic or genetically modified (GM) potato cultivars with excellent resistance to PLRV, PVY, Colorado potato beetle, and blackspot bruise. None of these GM cultivars is being grown currently, either in conventional or organic production (use of GM seed is prohibited in certified organic production). The ability to insert genes for resistance to pest problems, and thereby reduce the amount and types of pesticides used, has been quite successful in many other crops and holds great promise. More information on environmental impact and the risks and benefits of GM plants is needed to address public concerns about this methodology.

Cultural Practices

Proper management of the potato crop, from field preparation and planting through harvesting and storage, is essential for maximum yields of high-quality tubers. Many cultural practices,

Table 6. Relative Resistance of Potato Cultivars Grown in the Western United States to Common Diseases and Disorders.

Cultivar	common scab	net necrosis	early blight foliage	early blight tuber[2]	late blight foliage	late blight tuber	Verticillium wilt	blackleg	Potato virus X	Potato virus Y	growth cracks	hollow heart	second growth	storage decay	other[3]
Alturas	MR	MR	R	MS	S	MS	HR	MS	S	S[4]	S	R	R	MR	R to bb
Atlantic	MR	HR	S	MS	S	S	MR	MS	HR	S	MR	S	R	HS	S to ibs
Bannock Russet	R	R	MR	MR	S	S	R	MR	MR	HR	MR	S	R	MR	S to sb
CalWhite	MR	S	MS	MR	S	S	MS	MS	S	HS	MR	R	R	MR	MR to bb
Cherry Red	S	R	S	MS	S	S	MS	S	S	S	MR	R	R	S	
Chieftain	MR	R	S	MR	S	S	MS	MS	S	S	R	R	R	MR	S to o MR to bb
Chipeta	S	R	R	R	S	S	R	MS	S	MS	R	R	R	S	R to bb
Defender	S	MS	MS	MR	R	R	R	MS	R	MS	MR	MS	MR	MR	S to bb R to sb
Gem Russet	MR	R	MS	R	S	MS	S	MS	HR	HS	MR	MS	R	MR	S to bb
Norland	MS	R	S	MR	S	S	MS	S	S	MR	MR	R	R	MS	MR to bb
Ranger Russet	S	S	MR	MR	S	S	MR	MS	HR	R	MR	HR	MR	MR	S to bb
Red LaSoda	S	MS	MR	MS	HS	HS	MS	MS	S	—	MR	S	R	MR	S to ibs
Russet Burbank	R	HS	MS	MR	S	MS	MS	MS	S	S	MS	MS	HS	MR	S to bb
Russet Norkotah	R	S	S	R	S	HS	HS	MS	S	S	R	MS	R	MS	S to bb
Russet Norkotah strains	MR	S[5]	S	MR	S	S	MS	S	S	S[6]	R	MS	R	S	MR to bb
Russet Nugget	HR	MR	MR	MR	S	S	MR	MS	S	S	MR	R	R	MR	R to bb
Sangre	MS	R	S	MR	S	MR	MS	MS	MR	MR	MR	MR	R	MR	MR to bb
Shepody	HS	S	S	MR	MS	S	S	S	S	S	MR	MR	MR	S	MR to bb
Umatilla Russet	R	R	MS	MR	S	R	MR	S	S	MR	MR	MR	MR	S	S to bb
White Rose	MS	R	MS	MS	S	S	MR	MS	S	—	S	MR	MR	MS	MR to bb
Yukon Gold	S	MR	S	S	S	S	S	S	MR	S	MR	S	R	S	MS to bb

1. R = resistant, S = susceptible; H = highly, M = moderately, — = no information.
2. Immature tubers are more susceptible.
3. ibs = internal brown spot, bb = blackspot bruise, o = ozone, sb = shatter bruise.
4. Tubers susceptible to necrosis symptoms caused by tuber necrotic strains of PVY.
5. Some strains are resistant.
6. Susceptible, but infected plants yield better than infected plants of regular Russet Norkotah.

including seed selection and handling, planting, irrigation, fertilization, vinekill, and careful harvesting and storage methods have a significant impact on pest damage. Even when you cannot choose cultural methods solely for their effect on pest management, it is important to understand their impact on pests so that you will know what to expect.

Manage your potato crop so that tuber growth begins soon enough to produce adequate yield and remains uniform during the growing season. Careful water management early in the season helps prevent Rhizoctonia and seed piece decay, reduces the severity of Verticillium wilt and common scab, and helps prevent tuber malformations and tuber rots as plants mature and die. Early, excess fertilization of indeterminate cultivars delays tuber initiation and may reduce yields. Improper or inadequate fertilization also reduces yields and aggravates some diseases such as early blight and Verticillium wilt.

Sanitation. Some pest infestations begin when contaminated seed tubers, soil, water, or machinery is brought into a field. Take the following precautions to prevent the introduction or spread of potential problems.

- Use highest-quality certified seed tubers. Choosing high-quality seed may be the single most important decision one makes when growing potatoes. Root knot nematodes and damaging viral, bacterial, and fungal pathogens may be carried in contaminated tubers. Certification does not guarantee complete freedom from seedborne pests, but their spread is greatly reduced by using highest-quality certified seed. The extra expense is more than compensated by the reduction in pest problems, but remember that quality is not necessarily related to the cost.
- Thoroughly clean seed-cutting equipment with steam or pressurized water and then apply a suitable disinfec-

tant. Table 7 lists recommended disinfectants; be sure to keep surfaces moist with disinfectant for the recommended time and to rinse off corrosive disinfectants. Keep cutter blades sharp to reduce tissue tearing, which increases susceptibility to infections. Disinfect at least once a day and ALWAYS BETWEEN SEED LOTS.

- Do not irrigate with tailwater from fields known to be infested with root knot nematodes or other soilborne pathogens or pests; do not use water that may carry runoff containing harmful herbicides.
- If irrigation water comes from canals or ditches, consider installing screens to filter out seeds, rhizomes, and other weed parts.
- Clean equipment between fields to avoid moving contaminated soil.
- Schedule tillage to maximize the destruction of culls, harmful pests that survive or reproduce in tubers, and insects that pupate in the soil. In areas with cold winters, tilling after the first killing frost or leaving tubers on the soil surface through the winter usually destroys the most culls and leaves the fewest potato volunteers. Till immediately after harvest in warmer areas, being sure culls are buried to reduce tubermoth activity where this pest occurs. Rototilling is sometimes used when large numbers of culls are expected to survive.

- Make sure residue from previous crops has decayed before planting. Several pathogens survive between seasons on crop debris.
- Use roguing or spot treatments with herbicides to destroy problem weeds before they produce seed, especially dodder, nightshades, and perennials that are hard to control.
- Destroy weeds along field borders; if allowed to mature, they produce seeds that are a source of infestations in the cultivated area. A number of weeds can be reservoirs for the buildup of insect pests or pathogens that can subsequently spread to nearby potatoes.

Crop Rotation. Proper crop rotations enhance soil fertility, help maintain soil structure, reduce certain pest problems, increase soil organic matter, and conserve soil moisture. Herbicides not available for potatoes can be used in certain rotation crops to control problem weeds. Whenever possible, use rotations that reduce problem pests and avoid rotations that may increase them.

Generally, the most useful rotations for potato fields are forage crops and grains, including corn. Rotation crops that are useful only in some areas include alfalfa, cotton, sugar beets, beans, broccoli, and sudangrass. Grain and hay crops reduce Pacific Coast wireworm and northern root knot

Table 7. Disinfectants Commonly Recommended for Potato Handling Equipment and Storage Facilities.

Disinfectant	Effectiveness for		Inactivation by		Corrosiveness	Safety	Concentration	Exposure time	Shelf life	Comments
	Wet bacteria and slime	Dry bacteria and slime	Organic matter	Hard water						
Quaternary ammonium compounds	excellent	excellent	slightly	no	slight		label directions	10 min	1–2 yr	Diluted disinfectant relatively safe, concentrated form is poisonous. Stainless.
Hypochlorites (5.25% bleach)	excellent	excellent	yes	no (except iron)	yes	irritant, caustic	1:50 (0.1%) or 1:200	10 min	3-4 mo undiluted	Quick acting, inexpensive; caustic to skin and clothing. Use at 1:50 when mixing with water only. For maximum effectiveness use 1 part 5.25% bleach:200 parts water:0.6 parts white vinegar.
Iodine compounds	excellent	excellent	some; not greatly	no (except iron)	yes	relatively safe	label directions	10 min	1–2 yr	Do not take internally. No longer effective if it loses yellow-brown color. Tamed iodophor compounds work best.
Phenolic compounds	excellent	excellent	some; not greatly	no	no	oral poison	label directions	10 min	1–2 yr	Provides residual action. These have name "phenol" on label of ingredients.
Chlorine dioxide	excellent	excellent	slight	no	yes	concentrate and fumes can be toxic	label directions	10 min	2 weeks when mixed; 1–2 yr in separate containers	Activate prior to use. Broad-spectrum activity against viruses, fungi, and bacteria. Environmentally safe.

Source: Adapted from Guidelines for Seed Potato Selection, Handling, and Planting. Extension publication PP-877, North Dakota State Univ., 1997. REGISTRATIONS MAY VARY; CHECK WITH LOCAL AUTHORITIES.

nematode, but some increase wheat wireworm and Columbia root knot nematode. Alfalfa rotations reduce wireworms and Columbia root knot nematode, but most alfalfa cultivars increase northern root knot nematode. Broccoli rotations reduce the level of *Verticillium* in the soil. The benefit of a rotation crop may be increased by selecting cultivars of that crop that are more resistant to problem pests.

Green Manures. Incorporating a cover crop or the residue of a rotation crop as a green manure can reduce the inoculum levels and effects of some diseases and nematode pests, help reduce populations of problem weeds, and improve soil structure and fertility. Levels of Verticillium wilt are reduced by incorporating certain crops as green manures, including barley, mustard, rapeseed, sudangrass, and sweet corn. Populations of root knot nematodes may be reduced by incorporating oilseed radish, mustard, or sudangrass as green manure. The effect of green manures appears to be due in part to increased activity of beneficial microorganisms in the soil. These beneficial microbes may compete with pathogens, may affect pathogens directly, or may secrete substances that increase the resistance of the crop host. Some green manure crops produce chemicals toxic to pathogens when the crops break down in the soil. Mustard family plants release isothiocyanates; sorghum and sudangrass release hydrogen cyanide.

For more information on using green manures, see "Using Green Manures in Potato Cropping Systems" listed in the suggested reading.

Seed Tuber Handling. The longer a seed tuber is stored or aged after its dormant period is over, the more sprouts it produces, and consequently the more mainstems. Aging usually occurs during the time seed tubers are stored. Some aging is desirable to break the dormancy of enough eyes so that each seed piece will produce two or three stems. Cultivars such as Bannock Russet, Summit Russet, and Chipeta, which have a long dormancy period, must be aged longer than Russet Burbank to ensure a sufficient number of sprouts. However, if seed tubers are stored too long they give rise to less vigorous plants that produce many weak stems, smaller tubers, and lower yields.

Store seed tubers at 35° to 38°F (1.7° to 3.3°C). Approximately 2 weeks before cutting, warm seed tubers gradually to 50° to 55°F (10.0° to 12.8°C) and hold them at that temperature with a relative humidity greater than 90% and good ventilation. This reduces the amount of tissue tearing during cutting and encourages wound healing (suberization) after potatoes are cut, greatly reducing the incidence of seed piece decay after planting. If tubers are cut when they are just beginning to sprout, a stage sometimes called "peep" or "peek," emergence is more rapid and you can more easily choose a seed piece size that gives the number of sprouts you want. Cut seed tubers before

sprouts exceed about ⅛ inch (3 mm) in length to avoid breaking them and reduce the chance of spreading disease during cutting. If sprouts are broken, spread of mechanically transmitted viruses is more likely and seed pieces may develop multiple sprouts, which are weaker and may form too many stems per hill.

A seed piece size of 1.5 to 2.5 ounces (42.5 to 71 g) is recommended for optimum performance in most areas. The larger size is recommended for cultivars that have few eyes, such as Bannock Russet, CalWhite, and Russet Nugget. Cut seed in an area out of drafts. Follow good sanitation practices during cutting; clean and disinfect cutting equipment thoroughly between seed lots (see Table 7). Protect the cut seed from sun and wind when hauling.

Plant cut seed pieces immediately in moist soil (60% to 80% of field capacity) that is at a minimum of 45°F (7.2°C) to accelerate emergence and wound healing after planting. If you cannot plant cut seed immediately, hold it at 50° to 55°F (10.0° to 12.8°C) with good aeration and high humidity to speed wound healing. Store cut seed only where adequate airflow can be maintained throughout the pile. Do not store cut seed in bulk trucks. Do not plant seed that is cooler than the soil, particularly early in the season.

Seed Treatment. Seed treatments are used for a number of diseases, including Rhizoctonia, silver scurf, late blight, Fusarium dry rot, and bacterial soft rot. Materials containing a neonicotinoid insecticide can be applied to protect seed pieces from wireworm damage and to control sucking insects (aphids, psyllids, leafhoppers) and Colorado potato beetle for a substantial portion of the season.

When good-quality seed tubers are handled carefully and field conditions remain favorable for rapid wound healing and emergence, chemical seed treatments may not be necessary. Depending on the growing area and planting date, some growers can safely avoid the expense of seed treatments, while others may benefit from the insurance provided by applying them regularly. In areas where silver scurf, Rhizoctonia, or insect pests such as aphids or Colorado potato beetle are a regular problem, seed treatments may be applied routinely. If these pests are not a regular problem, and you are mostly concerned about seed piece decay, assess your needs for treatment by understanding the conditions that promote this disease. To make a decision, you must know about their frequency and severity in your situation. Important points to consider include:

- How confident can you be that soil moisture at planting will be optimal?
- Are the cultivars you plant resistant or susceptible to seed piece decay?
- How well controlled is your seed-cutting operation? Can you avoid conditions unfavorable to wound healing before planting?

- How predictable are the condition and quality of the seed tubers you use?
- What is your planting time? What is the likelihood of unfavorable postplanting weather conditions?

Planting. Proper field preparation is essential for maximum quality and yield. Residues from previous crops must be thoroughly incorporated to allow complete decomposition. Excessive undecomposed organic matter favors the development of common scab. Till finer-textured soils in the fall where possible; work sloping land as late as possible in fall or early in spring to avoid erosion. A winter cover crop is necessary in some areas to prevent erosion of coarse soils by wind; till these fields in the spring. If spring wind erosion is a problem, leave the crop residue on or near the soil surface.

Plow depth and equipment vary with location and conditions. Plow deep soils to a depth of 8 to 12 inches (20 to 30 cm); plow shallow soils no deeper than 1 inch (2.5 cm) below the plow sole. A plow width of 14 inches (35 cm) or more is satisfactory for potato fields. If a large quantity of residue must be turned under, use a plow with a 16- to 18-inch (40- to 45-cm) bottom and high clearance. Chiseling is recommended in many areas to break up compacted soil and ensure proper drainage. In areas such as the Columbia Basin where wind erosion is a potential problem, use shallow cultivations and keep as much crop residue as possible on the soil surface. Level the field so water can be applied uniformly without ponding. Basin or reservoir tillage, which uses a "dammer-diker" to leave dams spaced periodically within furrows, may be desirable on uneven land to make sure water penetrates evenly and does not run between hills. Beds prepared for potato planting are usually flat, but may also be raised. Preplant fertilizer and some soil-applied pesticides may be mixed into the soil at the time of bed preparation.

Plant seed pieces in soil that has a moisture level of 60% to 80% of field capacity from the level of seed pieces to 2 feet (0.6 m) below the surface. Apply a preplant irrigation if soil is dry. This results in a much higher and healthier plant stand than if irrigations are applied between planting and emergence. Be sure the soil temperature is at least 45°F (7.2°C) before planting. A soil temperature of 50° to 55°F (10.0° to 12.8°C) at planting depth is ideal for encouragement of vigorous sprout emergence, but in many areas waiting for the temperature to reach 50°F may mean losing valuable time. Planting in cold soil or soil that is either too wet or too dry increases the likelihood of damage by Rhizoctonia and seed piece decay, and therefore increases the need for seed treatments.

Planting depths vary from 4 to 6 inches (10 to 15 cm), depending on the area, time of year, cultivar, and conditions at planting. Seed pieces are planted deeper in coarse soils than in finer-textured soils. Deeper plantings are sometimes used for frost protection. Generally, shallower planting is used where emergence is expected to be slower. Be sure hilling operations give a final seed piece depth (from seed piece to the top of the ridge) of 6 to 8 inches (15 to 20 cm) by the time tubers begin to form.

Plant seed pieces in rows that are 30 to 36 inches (76 to 91 cm) wide; the width used is usually determined by local cropping patterns and equipment. Spacing of seed pieces within rows may control the number of tubers and tuber size. Potatoes grown for seed are often planted closer together to produce larger numbers of smaller tubers. In commercial production, spacing is influenced by cultivar. Russet Burbank is usually planted at spacings of 9 to 13 inches (23 to 33 cm). Spacings of 6 to 10 inches (15 to 25 cm) are used for cultivars that have few eyes per tuber or tend to produce fewer tubers per hill, and narrower spacing helps keep tubers from growing too large. Most red cultivars are planted 6 to 8 inches (15 to 20 cm) apart to favor the production of smaller tubers, which have a premium market price. For seed production, most cultivars are planted 6 to 8 inches apart. Crop management profiles have been developed for each new cultivar by the releasing university; these profiles include recommendations for planting depth and spacing. The quantities of seed tubers needed for different row widths and seed spacings are listed in Table 8.

Fertilizer applications are made preplant, at planting, or split between the two. Fertilizer bands are placed at least

Table 8. Number of Seed Pieces Used per 100 Feet of Row and Total Required per Acre for Different Planting Dimensions. Seed piece size 1.5 ounces.

Seed piece spacing inches (cm)	Number per 100 feet (per 30 m)	Total (cwt) required per acre for row widths Inches (cm)			
		30 (76)	32 (81)	34 (86)	36 (91)
6 (15)	200 (197)	32.6	30.7	28.8	27.1
8 (20)	150 (148)	24.5	22.9	21.6	20.4
10 (25)	120 (118)	19.6	18.4	17.3	16.3
12 (30)	100 (98)	16.3	15.4	14.4	13.5
14 (35)	86 (85)	14.0	13.1	12.4	11.6

Metric conversion: 1 cwt = 45.36 kg.

2 inches (5 cm) to the sides and 2 inches above or below the seed pieces, depending on irrigation method—above for overhead irrigation, below for furrow irrigation. Otherwise, seed pieces may be damaged by high salt concentrations, causing stand reductions. Pick and cup planters are the most popular types used for potatoes (Figs. 13 and 14). Keep the planter in good operating condition and check to be sure it is adjusted to work properly with the seed size you are planting. Proper ground speed is critical for proper operation and accuracy of all planters. Check the accuracy of seed placement by digging up sections of each planter row and measuring actual versus desired placement. With a cup planter, use cups that are the correct size; with a pick planter, use pick lengths and an arrangement suited to the seed size. To reduce the risk of disease spread, avoid using pick planters for planting seed potato crops.

When plants begin to emerge, count the number in 100-foot sections and compare to the expected number for the spacing you used (see Table 8). If the number of emerged plants is significantly lower than expected, dig up seed pieces to determine the cause of the reduced stand. Poor stands may be caused by mechanical errors in planting, blank ("blind") seed pieces (lacking eyes), exposure of seed to sprout inhibitors, herbicide residue in soil or seed, disease, or damage during cultivation.

Hilling. Perform hilling operations after plants begin to emerge and do not cover the emerging plants. Discs, rolling cultivators, hilling listers, or winged cultivators are used. Hilling allows the use of a shallow seed piece planting depth for rapid emergence while providing the soil depth necessary later in the season for proper tuber development and reduced exposure to sunlight and adverse temperatures. Exposed tubers develop green color due to chlorophyll production and bitter-tasting compounds called glycoalkaloids in the skin and cortex; they are also susceptible to tuberworm infestation in the Columbia Basin, Sacramento–San Joaquin Delta, and low-elevation desert valley growing areas of California. Shallow tubers are more susceptible to heat necrosis. For best protection of tubers, be sure hills are flat and broad rather than narrow and peaked. Cultivation during hilling destroys emerged weeds, and some soil-applied herbicides and insecticides can be mixed in at this time. When applying herbicides during hilling operations, be sure not to cover sprouts with treated soil; set equipment to throw treated soil back into the furrow. Plan to complete hilling before plants are so large that pruning of stolons or roots may occur.

Drag-off is the practice of removing the tops of hills to hasten emergence, usually when seed pieces have been planted in raised beds. It is not widely used and must be performed carefully because of the potential for damaging sprouts.

Frost Protection. In growing areas where frost may occur after plants are up, irrigation can keep the foliage from

Figure 13. A pick planter uses metal spikes or picks, available in various lengths, to transfer seed pieces from the bin into the seedbed. Picks can spread pathogens from infected to uninfected seed pieces.

Figure 14. A cup planter uses cups, available in different sizes, to transfer seed pieces into the seedbed. Both pick and cup planters can be used to apply fertilizer during planting.

Soil is about half solid material by volume (large circle). The rest of the soil volume consists of pore spaces between soil particles. Pore spaces hold varying proportions of air and water (small circle).

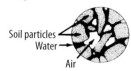

When the soil is **saturated** after irrigation or rain, pore spaces are filled with water.

When soil has drained following irrigation, it is at **field capacity**. In most soils, about half of the pore space is filled with water. About half of this water is **available** to plants; the rest is unavailable because too much suction is needed to remove it from pore spaces.

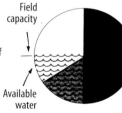

The **allowable depletion** is the proportion of the available water the crop can use before irrigation is needed.

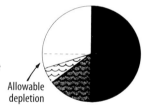

At the wilting point, all available water is gone. Plants die unless water is added.

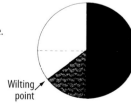

Figure 15. The soil reservoir.

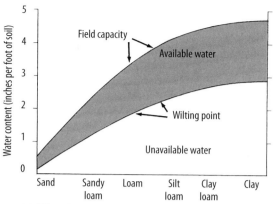

Figure 16. The relationship between soil texture and water-holding capacity.

freezing. Solid set sprinklers are usually required; they must be turned on before the air temperature drops below freezing and left on until the air temperature has risen and the threat of freezing has passed. As water freezes, it releases a small amount of heat, which is sufficient to keep the leaf tissue from freezing. However, new drops of water must freeze continuously throughout the frost period to keep foliage above the freezing point. Using irrigation for frost protection increases the risk of some diseases. In some locations, helicopters may break up layers of cold air at ground level and keep foliage from freezing.

Irrigation. Availability of soil water is a major factor that determines yield and quality of the potato crop. Too little water reduces yields, induces tuber malformations, or increases severity of common scab or Verticillium wilt after infection has occurred. Excess or poorly timed irrigation may reduce yields and quality, cause several disease problems in the field or in storage, or leach nutrients from the root zone. Fluctuations in water availability favor disorders such as second growth and internal necrosis.

Efficient irrigation requires finding out how much available water the soil can hold. Available water is that portion of the soil water that can be withdrawn by plants (Fig. 15). During the growing season, irrigation is needed when a certain proportion of the available water, the allowable depletion, has been used. The allowable depletion in a particular field varies according to soil type, stage of crop growth, total amount of available water, weather conditions, and irrigation cost. Because potatoes are sensitive to water stress, the allowable depletion is no more than 30% to 40%. To minimize common scab infection, the allowable depletion is no more than 20% during tuber initiation.

Potatoes are a shallow-rooted crop; 90% of the roots grow in the top 12 to 18 inches (30 to 45 cm) of the soil. You can determine from the clay content and soil texture in the top 18 inches how much available water the soil can hold (Fig. 16). For a soil profile that includes layers of different soil types, calculate the available water separately for each layer; then add them together to obtain the total available water in the rooting zone (Fig. 17). Take salinity into account in estimating available water; if the soil or irrigation water contains high levels of salts, plants withdraw less water, so the available water will be less than that estimated using Figure 17.

Irrigation Methods. Most potatoes in the western states are irrigated with sprinklers. Center pivot, wheel line, and solid set systems are most commonly used. Sprinkler systems are more versatile than furrow irrigation systems and can apply fertilizers and some pesticides effectively. Uniform water application is most easily achieved with sprinkler systems. Sprinklers are readily adapted to uneven ground. When preparing fields, be sure not to leave any low spots where water will collect. Sprinkler irrigation provides conditions in the potato canopy that are favorable for certain diseases, such as early blight, late blight, bacterial stem rot, and white mold (Sclerotinia). To reduce spread of these diseases, allow foliage to dry between irrigations. An advantage of center pivot irrigation systems is that the water applied is added relatively quickly to the plants. However, regardless of irrigation technique, watering during late afternoon or early evening may allow foliage to stay wet all night, providing a favorable environment for late blight. Allowing solid set sprinklers to run more than 6 hours, regardless of the time of day, has the same effect. Also, foliage near the middle of center pivot circles tends to remain wetter and more prone

to foliar diseases and produces poorer quality potatoes, so this area should not be planted.

Drip irrigation systems are the most efficient, typically requiring 10% to 20% less water than sprinklers. The risk of foliar diseases is lower with drip systems, and they can apply fertilizers and some pesticides effectively. However, drip systems may create challenges with tillage and harvest operations and are particularly challenging when used on very coarse soils. Mite infestations may increase with the use of drip irrigation. The expense of drip systems generally makes them uneconomical for commercial potato production.

Furrow irrigation may be used where the slope of a field is less than 2%, rows are not longer than 600 to 800 feet (182 to 244 m), and the soil does not have an excessively high infiltration rate. The length of furrows is affected by soil texture. Uniform application of water and fertilizer is more difficult with furrow systems. Tuberworm damage is almost always higher where furrow irrigation is used within the range of this pest (Columbia Basin, Sacramento–San Joaquin Delta, and lower desert valleys of California) because soil cracking is more prevalent. Powdery mildew is favored by furrow irrigation, translucent end occurs more frequently, and the potential for Verticillium wilt is higher than with sprinkler irrigation. Early blight and late blight are less favored by furrow irrigation.

Subirrigation can be used where the water table can be raised easily, soil is of uniform texture and structure, and fields are relatively level. The low, flat, peat soils of the Sacramento–San Joaquin Delta area of California are favorable for the use of subirrigation, but the lack of a high degree of soil uniformity makes other irrigation systems more advantageous.

Preirrigation. Soil moisture should be at 60% to 80% of field capacity at planting time. If rainfall is not adequate to fill the soil reservoir, use a fall irrigation or irrigate before planting. Avoid irrigations between planting and emergence. Irrigations at this time can increase blackleg, Rhizoctonia stem and stolon canker, and seed piece decay. It is best to have soil moisture high enough so that the first irrigation is not needed until plants have emerged. However, irrigations must be used if the soil becomes excessively dry, and they may also be needed to reduce wind erosion.

Postplant Irrigations. With sprinklers, each postplant irrigation should bring the top 18 inches (45 cm) of soil back to field capacity; do not irrigate to a depth of more than 24 inches (60 cm). The timing and amounts of postplant irrigations depend on the water-holding capacity of the upper 18 inches of soil and the rate at which the water is used by evapotranspiration. Maintaining adequate soil moisture is critical during the tuber initiation and tuber growth phases; water stress during these periods may cause tuber malforma-

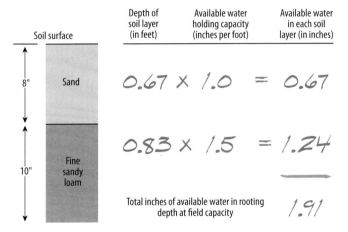

Figure 17. To calculate how much water the soil can hold at field capacity, first examine the soil profile to see if there are layers of different soil textures in the rooting depth of the crop. For each soil layer, multiply the depth of the soil layer in feet (or meters) by the available water-holding capacity of that soil texture (see below). Add the results from all the layers to get the total available water-holding capacity for the soil profile.

Sand	1.0 inches/foot (8.3 cm/m)
Sandy Loam	1.5 inches/foot (12.5 cm/m)
Loam, Silt Loam, Clay	2.2 inches/foot (18.3 cm/m)

Table 9. Judging Depletion of Soil Water by Feel and Appearance.

Coarse-textured soils	Inches (mm) of water needed[1]	Medium-textured soils	Inches (mm) of water needed[1]	Fine-textured soils	Inches (mm) of water needed[1]
Soil looks and feels moist, forms a cast or ball, and stains hand.	0.0 (0.0)	Soil dark, feels smooth, and ribbons out between fingers; leaves wet outline on hand.	0.0 (0.0)	Soil dark, may feel sticky, stains hand; ribbons easily when squeezed and forms a good ball.	0.0 (0.0)
Soil dark, stains hand slightly; forms a weak ball when squeezed.	0.3 (7.5)	Soil dark, feels slick, stains hand; works easily and forms ball or cast.	0.5 (12.5)	Soil dark, feels slick, stains hand; ribbons easily and forms a good ball.	0.7 (17.5)
Soil forms a fragile cast when squeezed.	0.6 (10.5)	Soil crumbly but may form a weak cast when squeezed.	1.0 (25)	Soil crumbly but pliable; forms cast or ball, will ribbon; stains hand slightly.	1.4 (35)
Soil dry, loose, crumbly.	1.0 (25)	Soil crumbly, powdery; barely keeps shape when squeezed.	1.5 (37.5)	Soil hard, firm, cracked; too stiff to work or ribbon.	2.0 (50)

1. Amount needed to restore 1 foot of soil depth to field capacity when soil is in the condition indicated.

tions and translucent end, especially in Russet Burbank. Dry conditions before tuber initiation discourage Verticillium wilt but favor common scab infections; be sure to know which disease is of greater importance in your area. Overly wet soils favor powdery scab infections.

Final irrigations in most growing areas should be timed to allow soil moisture to drop to about 60% of field capacity at the time of vinekill. This level of soil moisture encourages proper development of the tuber skin and decreases the chance that tubers will be infected during harvest by early blight, late blight, or soft rot pathogens. However, if soil moisture falls below 50% during vinekill, stem end browning may result. In hot growing areas, light irrigations may be continued until harvest to keep soil temperatures down. Excess irrigation at this time may reduce oxygen levels in the soil and cause tuber rot or black heart.

Irrigation Scheduling. Timing of irrigations during the growing season can be based on various measures of soil moisture together with a water budget, which can be based on evapotranspiration or pan evaporation data. A combination of methods is usually best, and disease potential must be considered.

Always check soil moisture before applying water and estimate how much available water remains in the crop rooting depth. Use a soil tube or shovel to take soil from the rooting zone at several points in each field. In furrow-irrigated fields, check a few places at the head end of the runs, some in the middle, and some at the tail end. Use Table 9 as a guide for judging the depletion level in soil taken from the root zone.

Water needs usually vary from one part of a field to another, especially if the field includes different soil types or slopes. Plants in a sandy streak or where root growth has been restricted show stress sooner than the rest of the crop. Watch these weak areas to gain advance notice of when irrigation is needed for the rest of the field. However, schedule irrigations according to the need shown by most of the crop.

Instruments are available that measure the moisture content of the soil. Tensiometers and neutron probes are frequently used for monitoring soil moisture. To obtain reliable readings, you must install these instruments in areas representative of the field, including spots where water stress occurs more readily. Aerial infrared photography can help identify areas of different moisture stress within fields. At each site install one soil moisture probe at the rooting depth of the current growth stage and a second probe 18 to 24 inches (45 to 60 cm) deep. Follow the recommendations of irrigation experts in using soil probes for irrigation scheduling.

A convenient way to monitor soil moisture indirectly is to use a water budget (Fig. 18) to estimate how much water the crop uses from day to day under prevailing weather conditions. After soil has drained to field capacity, further loss of soil water occurs mainly through evaporation from the soil surface or transpiration from leaves. The combination of evaporation and transpiration is called evapotranspiration (ET). If you know how much available water is in the crop rooting depth at field capacity and how much water is lost through ET each day, you can estimate the amount of available water remaining at any time by adding up the daily ET values. Newspapers and radio and television stations in some areas report ET figures for major crops, including potatoes. Water use information for most potato-growing areas is available on the Internet, from CIMIS for California locations (www.cimis.water.ca.gov) and AgriMet for locations in the Pacific Northwest (www.usbr.gov/pn/agrimet). Information is also available from private consulting firms; check with local authorities regarding sources of information for your area.

You can also use daily pan evaporation and a water use curve for the crop to develop your own ET figures. Figure 19 shows a water use curve for potatoes, representing water use at different stages of growth. To use this curve, determine the water use coefficient for the current stage of potato growth, for example, a value of 0.45 during tuber initiation. Multiply the daily pan evaporation by this figure to get actual crop water use. Pan evaporation can be obtained from local weather stations as well as the CIMIS and AgriMet Web sites.

Water budgets provide estimates of crop water use. However, actual water use is affected by cultivar, disease, weeds, insects, physical characteristics of individual fields, and management factors. Use a direct measurement of soil moisture to make the final decision about when to irrigate.

A good irrigation program follows these basic guidelines.

- Start with soil at field capacity and avoid irrigation between planting and emergence.
- Monitor the soil moisture.
- Record daily ET.
- Keep records of irrigation and rainfall amounts.
- Do not irrigate deeper than 18 to 24 inches (45 to 60 cm) or more frequently than necessary.
- If late blight is a risk, irrigate only when foliage will dry before nightfall; begin irrigation during early morning only when foliage will dry within 6 to 8 hours; do not run solid set sprinklers longer than 6 hours, regardless of time of day. Turn off the sprinklers near the center of a pivot as much as possible.
- Avoid wet soil (soil with moisture in excess of levels recommended for the current stage of growth.)
- Use irrigation systems designed to give uniform water distribution.
- Be sure to reduce irrigations, matching the decreased plant demand in late season (see Fig. 19), to avoid excess water that encourages development and spread of late blight as well as other tuber rots.

Managing Salinity. Prevention of harmful salt buildup is an important part of irrigation. As part of routine soil analysis,

FIELD NO. _____ 14

ALLOWABLE DEPLETION 40% of 2.0 in. = 0.80 inch

DATE	ET IN INCHES PER DAY	CUMULATIVE ET	
7/10			irrigation
7/11	0.23	0.23	
7/12	0.25	0.48	
7/13	0.27	0.75	check field
7/14	0.26		irrigation
7/15	0.27	0.27	
7/16	0.25	0.52	
7/17	0.24	0.76	check field
7/18	0.26		irrigation
7/19	0.28	0.28	
7/20	0.27	0.55	
7/21	0.25		irrigation

Figure 18. If you have a source of evapotranspiration (ET) data, you can use a water budget to estimate the interval between irrigations. You can also use pan evaporation and the crop water use coefficient (see Fig. 19 and text) to develop your own ET figures. Keep a running total starting the day after irrigation. When the total approaches the allowable depletion, check soil moisture in the field to make a final decision on when to irrigate.

have a laboratory determine the concentration of total salts and the proportion of sodium in the soil. A soil salinity reading of 2 to 5 mmho/cm indicates a possible salinity problem. If the salt concentration in the soil is high, plan to leach the salt below the root zone with extra water during preirrigation. Consult your local extension agent, farm advisor, or other experts about guidelines for your situation.

Test irrigation water for total salts and sodium every 2 or 3 years, especially if it comes from wells where the water table is dropping. Test each well separately, as salt concentrations may differ greatly, even among wells in the same area.

Fertilization. Continuous availability of adequate nutrients throughout the growing season is necessary for the best yield and quality and disease control. If nutrient deficiencies occur during tuber growth, the plant shunts nutrients from

Table 10. Approximate Amounts of Mineral Nutrients Removed from the Soil by Potato Vines and Tubers.

Nutrient		Amount removed (lb/acre) by different yields (cwt/acre)					
	Vines	300	400	500	600	700	800
nitrogen	139	128	171	214	257	300	343
phosphorus	11	17	23	29	35	41	47
potassium	275	144	192	240	288	336	384
calcium	43	4.4	5.9	7.4	8.9	10.4	11.9
magnesium	25	8.9	11.8	14.7	17.6	20.5	23.4
sodium	2.70	1.74	2.32	2.90	3.48	4.06	4.64
zinc	0.11	0.11	0.14	0.18	0.22	0.26	0.30
manganese	0.17	0.04	0.06	0.07	0.08	0.09	0.10
iron	2.21	0.79	1.06	1.32	1.58	1.84	2.10
copper	0.03	0.06	0.08	0.10	0.12	0.14	0.16
boron	0.14	0.04	0.05	0.06	0.07	0.08	0.09

Metric conversions: 1 lb/acre = 1.12 kg/ha; 1 cwt/acre = 0.1121 t/ha.

the roots, stems, and leaves to the growing tubers, thereby hastening aging of the vines. As a result, certain diseases such as early blight and Verticillium wilt are aggravated, and yields are reduced. On the other hand, excess fertilizer delays the onset of tuber growth and may reduce yields; tuber decay after harvest may also be increased, and processing qualities such as specific gravity may be lowered.

Soil nutrients removed by potato plants are directly related to yield (Table 10); that is, a yield of 600 hundredweight per acre (67.3 t/ha) removes about twice as much of the soil nutrients as 300 hundredweight per acre (33.6 t/ha). Nutrients used in the smallest quantities (zinc, manganese, iron, copper, and boron) are referred to as micronutrients. The need to add nutrients is determined by whether the soil can provide them in adequate amounts at the time the plants need them for optimum growth.

Nitrogen is almost always needed; phosphorus and potassium are frequently needed. Calcium and magnesium are needed on some soils. Zinc and manganese may be required on some calcareous, alkaline soils; they usually are applied as chelated micronutrients to prevent them from being tied up by chemical reactions in alkaline soil. In deficient, nonalkaline soils, they may be applied as zinc sulfate or manganese sulfate. Sulfur supplements are required in areas where the soil or water is low in sulfur or to help prevent common scab.

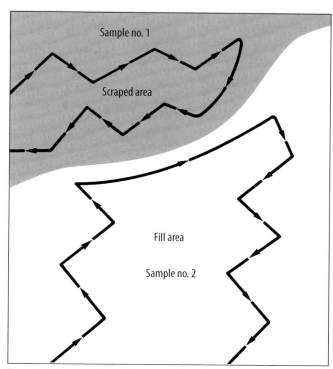

Figure 20. Suggested sampling pattern for taking petioles for nutrient analysis from a nonuniform field.

Muck or peat soils may benefit from additions of copper and manganese.

To determine fertilizer needs, consider cropping history and individual field history, records of petiole analysis from previous seasons, preplant soil analysis, and petiole and soil analysis throughout the season. For nutrients that can be added during potato growth (e.g. nitrogen), petiole analysis is very useful for determining when additions are needed. Petiole analysis also provides a useful guide for the next season, by indicating which nutrients are being depleted. Soil tests help determine the need for preplant and in-season applications of nitrogen, phosphorus, potassium, sulfur, and other nutrients. Sampling techniques and fertilizer recommendations are presented in several references listed in the suggested reading.

Fertilizer is used most efficiently when supplied closest to the time it is needed. This is especially true of nitrogen, which can be lost readily by leaching or microbial activity. More nutrients are generally needed in early and middle growth stages to assure maximum growth of vines, roots, and tubers and to assure adequate tuber initiation. Late application of nitrogen delays tuber growth and maturity. If drip irrigation is used, most nutrients can be added during the season because water and nutrients are added directly to the root zone. This is generally more efficient than adding all fertilizer preplant or at planting. Because growing conditions and cultivars differ from one area to the next, guidelines for

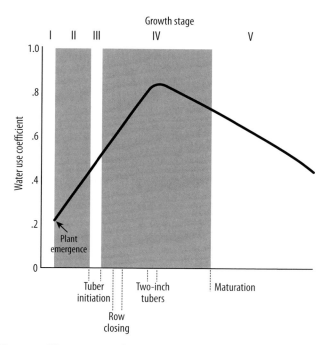

Figure 19. Water use curve for potatoes, adapted from *Crop Water Use Curves for Irrigation Scheduling*, Special Report 706, Agricultural Experiment Station, Oregon State University, Corvallis, OR. Daily water use (ET) can be calculated by multiplying the daily pan evaporation (available from newspapers or weather stations) by the water use coefficient for the current crop stage as indicated in this figure. These ET figures can be used in a water budget such as the one shown in Figure 18.

adequate nutrient ranges differ. Consult with local experts to get the latest guidelines for your area.

Most nutrients can be applied before planting in sufficient quantities to last all season. Phosphorus, potassium, and most micronutrients do not move readily with soil water; they are most often applied at or before planting. Whether fertilizer is applied as a preplant broadcast or banded at planting depends on soil type, soil temperature, and fertilizer formulation and cost. Banding near the seed pieces assures rapid availability and uptake under adverse conditions; broadcasting distributes the nutrients more completely throughout the root zone.

Because nitrogen is easily leached from the rooting zone, especially in coarse-textured soils, it is best to apply only a portion of the total requirement before planting. This is particularly important in areas with long growing seasons or soils where nitrogen is easily leached. The amount applied before planting depends on cultivar, soil type, and growing conditions. If leaching is not a problem, most or all of the seasonal requirement may be applied before or during the planting of some determinate cultivars. If equipped to apply fertilizer through irrigation water, apply no more than half the total nitrogen requirement before planting Russet Burbank or other indeterminate cultivars; otherwise, tuber growth will be delayed and yields may be reduced. Use petiole analysis to schedule application of the remainder of the seasonal requirement.

Soil Testing. A preplant soil test is the best way to determine what nutrients should be applied before planting. Consult references in the suggested reading about soil sampling, and follow the recommendations of local experts and the testing laboratory for taking soil samples. The best time to sample is between crops. Be sure to take samples that represent any differences in soil conditions or cropping history within a field. Soil tests can also indicate whether there is a need for calcium, manganese, sulfur, or any other micronutrient.

Precision agriculture uses selective or grid soil sampling methods to identify soil variability zones within a field. Zones with similar properties receive inputs formulated to correct deficiencies or deal with other issues on a site-specific basis. Consult the references in the suggested reading for specific information or check with local experts to get the latest guidelines for your area.

Soil test results for nitrogen are not as accurate as for the other nutrients because the amount of nitrogen available to plants can change rapidly. Some areas do not recommend basing seasonal nitrogen applications on soil test results. In-season soil sampling is used by some producers and consultants to monitor the soil's supplying power. These results are used to augment petiole analysis information. For specific fertilizer recommendations based on soil test results, consult the references appropriate for your area.

Petiole Analysis. The crop nutrient status during the season can be assessed by measuring nutrient concentrations in a portion of the vines. The most reliable way to do this is to take adequate and representative samples of petioles and have a reliable laboratory in your area analyze them for nutrient concentrations. Take the first sample when tuber initiation begins; in Russet Burbank this is about the time plants are at the 8-leaf stage and about 10 inches (25 cm) tall. Follow a pattern like the one illustrated in Figure 20, removing one petiole from each plant sampled. Always take the first fully expanded leaf below the top, usually the fourth leaf (Fig. 21). Remove the leaflets from each petiole and store all the petioles for one sample in a clearly marked paper bag. Good results are obtained with 40 petioles per sample and at least three samples per field. Deliver samples to the laboratory immediately; if you cannot, dry them at a temperature

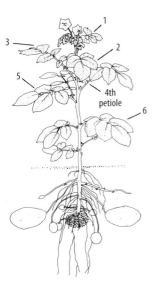

Figure 21. When taking petiole samples for nutrient analysis, select the fourth petiole (the first fully expanded leaf) from the growing tip (top). Remove all leaflets from the petiole (bottom) before sending samples to be analyzed.

of 100° to 150°F (38° to 66°C) or store them in a freezer until you can. Take at least three or four samples between tuber initiation and early maturation. Some areas recommend sampling at least once every 7 to 14 days. Keep records of all the results; they will be helpful in planning fertilizer applications for next season. Recommendations for adequate concentrations of major nutrients (nitrogen, phosphorous, potassium) based on petiole analysis are available for most areas, and recommendations for micronutrient concentrations are also available for some areas. Recommendations based on whole-leaf analysis are available in some areas. Check with local authorities and consult the publications for your area listed in the suggested reading.

Use petiole test results to schedule midseason nitrogen applications. Petiole nitrogen concentrations in Russet Burbank should be allowed to decrease beginning at least 3 weeks before vinekill. Otherwise, tubers will not mature properly, bruising and decay losses may be increased, specific gravity may be low, and sugar content may be high.

Guidelines for adequate midseason petiole nitrogen vary with area, application method, and cultivar. In some cases the market destination of the crop (i.e., chipping, french fries, fresh market, seed) affects fertilizer guidelines. Figure 22 shows examples of recommendations for Russet Burbank in Idaho and the Columbia Basin. Contact local experts to learn the recommendations for your area and the cultivars you are growing, and consult publications listed in the suggested reading.

Apply nitrogen through sprinklers as urea, ammonium, ammonium nitrate, or combinations of these. Use caution when irrigating; overirrigation may leach nitrate from the root zone. Make the first application at the time of tuber initiation, then use the trend shown by petiole analysis to schedule subsequent applications to keep the petiole level adequate. Making the final application about 4 weeks before vinekill usually allows nitrogen to decrease properly. In the Columbia Basin, final application of nitrogen may need to be as much as 6 weeks before vinekill to allow for proper maturation. However, in some soils petiole nitrogen decreases more quickly, and an additional nitrogen application may be needed. Continue to monitor petiole nitrogen after the final fertilizer application so that you will know if additional treatment may be necessary.

Petiole analysis may also indicate whether phosphorus levels are likely to become deficient during the season. Research in Idaho indicates that Russet Burbank yields are improved if petiole phosphorus is kept above 1,000 parts per million until 20 to 30 days before vinekill. Because petiole phosphorus decreases during the season, a graph can predict whether petiole phosphorus is going to drop below the critical level too soon (Fig. 23). If so, applications of a highly soluble form of phosphorus, such as ammonium polyphosphate, are recommended in some states; some other areas may not

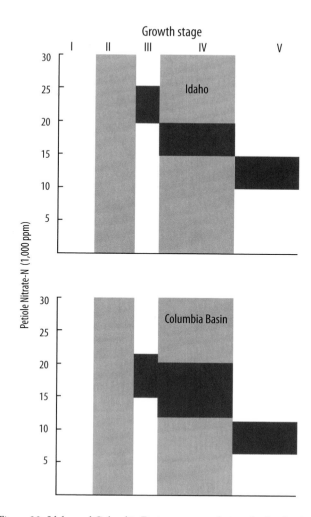

Figure 22. Idaho and Columbia Basin recommendations for levels of petiole nitrate nitrogen considered adequate for Russet Burbank potatoes at different growth stages: III = tuber initiation, IV = tuber growth, V = maturation. Brown-shaded areas show adequate levels.

recommend midseason phosphorus applications. The most widely recommended way to keep phosphorus above the critical level throughout the season is to apply adequate phosphorus before or during planting.

Potassium concentrations in petioles should not fall below the critical level (about 7%) during the season and should not decline rapidly. Rapid decline can indicate poor root growth caused by poor soil structure and poor soil aeration, or low soil potassium availability. Low potassium concentrations increase the susceptibility of tubers to bruising and the plant to some diseases. Midseason potassium applications can correct this problem effectively. Drip irrigation systems commonly include frequent applications of nitrogen and potassium during the growing season.

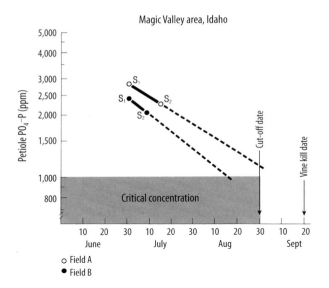

Figure 23. Two examples illustrating Idaho guidelines for estimating future petiole phosphate-phosphorus concentrations. Phosphate concentrations from two initial petiole samples (S1 and S2) are plotted on a semilogarithmic graph. The broken lines are extrapolated by drawing straight lines between S1 and S2; they predict a drop below the critical concentration (1,000 ppm) in field B before the cutoff date. Additional phosphorus fertilizer would be needed in field B, but not in field A.

Vinekill. In some areas, particularly in the case of early crops that will not be stored long-term but go directly to processors, potatoes are harvested while the vines are still green. In most cases, potatoes are harvested after the vines are dead. Harvesting after vine death provides several benefits. Tubers are allowed to mature and develop a thicker skin (periderm) so they are less susceptible to bruising and diseases. Specific gravity and sugars reach more desirable levels, which is especially important for potatoes to be processed. Inoculum of some diseases such as late blight is reduced by allowing vines to dry up before harvest. In the case of seed crops, tuber size can be controlled and virus-transmitting aphid populations, which usually are higher later in the season, can be avoided by killing vines early. Vinekill in early-season potato crops should be accomplished by the quickest method possible; when vines decline slowly aphids are more likely to develop winged forms, which may migrate to later-season cultivars, spreading viruses they may have acquired from the early crop.

Vine death occurs naturally in areas with early frosts and in hot areas where cutting back on water rapidly kills vines. In most areas, mechanical or chemical methods of vinekill are used. Rapid vinekill may cause vascular discoloration in tubers. It is usually confined to the stem end but sometimes spreads throughout the tuber. Vascular discoloration is more likely to occur if vine-killing agents are applied when soil moisture is below 50% of field capacity.

Follow these guidelines to reduce the risk of vascular discoloration and increase the effectiveness of vine-killing treatments.
- Avoid using chemical vinekillers during cool, wet weather.
- Avoid uncovering or damaging tubers during mechanical operations.
- Make sure soil moisture is adequate (at least 50% of field capacity).

Start vinekill 2 or 3 weeks before harvest. Timing depends on cultivar, growing conditions, time of year, and potentially the incidence and severity of late blight in the field. Lush vine growth takes longer to kill; split applications of vine-killing agents may be beneficial.

Harvest. Most potatoes are harvested by harvesters that dig two or more rows at a time and load directly into bulk trucks, which carry the tubers to a packinghouse, processor, or storage facility. Where soil is free of rocks and clods, windrowers are frequently used, allowing the harvest of up to twelve rows at a time.

Prevention of bruising is one of the most important considerations in a well-managed harvest operation. Blackspot and shatter bruise can seriously affect marketable yield if precautions are not taken to reduce them. Several factors are important in controlling bruising.

Soil Moisture. Proper soil moisture at harvest helps reduce bruising. The water content of tubers affects their sensitivity to bruising. If their water content is low, they are sensitive to blackspot bruise; if it is high they are sensitive to shatter bruise. The right combination of moisture and temperature helps to minimize both kinds of bruising. Soil moisture of 60% to 80% of field capacity is generally recommended.

Soil Temperature. The lower the soil temperature, the more susceptible tubers are to shatter bruising, especially when the temperature is below 50°F (10.0°C). Whenever possible, harvest when the soil temperature is above 50°F. Soils are warmest between 11 a.m. and 11 p.m. Harvest during these hours if soil temperature is likely to fall below 50°F at night. However, avoid harvesting when tuber pulp temperatures are above 60°F (15.6°C) to reduce tuber rots in storage.

Equipment Operation. Follow the recommendations for harvester operation in your area to reduce bruise injury. Use equipment that is in good repair, correctly adjusted, and operated by an experienced person familiar with local conditions. Important considerations are digger blade depth; reduction of "spill-out" losses at the digger blade; apron pitch, speed, and agitation; travel speed; drop heights; and control of "undersweep." Operate the primary chain at 1.1 times the travel speed (1.5 feet per second for each 1 mile per hour of tractor speed, or 0.31 meter per second for each 1 kilometer per hour of tractor speed). Operate the other harvester chains at 0.6 to 0.8 times the travel speed (1 to 1.2 feet per second for each 1 mile per hour; 0.17 to 0.22 meter per second for each 1 kilometer per hour). Keep chains filled so the tubers have little space to bounce around. Use padding on equipment wherever bruising might occur. When unloading at storage facilities, keep drop heights to a minimum; use straw-filled bags on baffle boards and use mats underneath hoppers to cushion spilled tubers.

Instruments that measure bruising forces in harvesting and handling equipment can be used to correlate the measured forces with resulting bruising, and thereby identify specific locations where modifications are needed to reduce bruising.

Sprout Inhibitors. Apply sprout inhibitors to fresh market or processing potato tubers that are to be stored for more than 2 or 3 months. Low storage temperature cannot prevent sprouting without undesirable accumulation of sugars. Two types of sprout inhibitors are used: one, maleic hydrazide, is applied to potatoes while they are still actively growing; the other, chlorpropham (CIPC), is applied through the ventilation system in storage. Never apply these sprout inhibitors to seed potatoes or put seed potatoes in a storage facility that has been treated with CIPC while awaiting cutting and planting.

If you are using maleic hydrazide, the timing of application is critical. If applied too early, yields are reduced and injury, including premature crop senescence, may result; if applied too late, the material will not be translocated into the tubers and sprouting will not be inhibited. Apply maleic hydrazide when all the tubers you intend to market are at least 1½ to 2 inches (4 to 5 cm) in diameter, when the plants are in the third (tertiary) bloom, or when lower leaves turn yellow. Do not apply the chemical to vines that are stressed by low water, frost, or disease. Some maleic hydrazide labels prohibit its use on irrigated potatoes in the western states;

check the product label to be sure use is allowed on the potatoes you produce.

Chlorpropham (CIPC) can be applied to stored tubers either as an aerosol spray or through the ventilation system. Apply CIPC after the curing period; use specially trained operators and follow label directions regarding storage management after application. Avoid using CIPC in storage areas that are not equipped with forced-air ventilation. Avoid handling tubers after treatment with chlorpropham; they do not heal (suberize) easily. Avoid storing seed tubers in a building where tubers have been treated with CIPC. Avoid using CIPC on lots with pockets of soil that will interfere with airflow or on lots that have already started to sprout.

Storage. A large part of the crop in many growing areas is stored for fresh market or processing during the winter and spring. All seed tubers are stored. Designs can vary, but most storage facilities have controls for temperature, humidity, and ventilation. Ventilation is essential during storage. It removes field heat, excess moisture that may condense on colder tubers, and carbon dioxide and heat produced by respiration; at the same time it helps provide even temperature and humidity within the storage area and oxygen to support tuber respiration. Uniform airflow throughout the pile is important. For storage requirements in your area, consult local experts and publications listed in the suggested reading.

To reduce the risk of rot developing and spreading during the storage season, have wet or rotting tubers removed from the incoming conveyors when filling a storage unit. Don't mix "good" and "bad" lots in the same storage, and place lots with possible problems nearest the door so they can be removed without impacting other lots in the storage.

The storage period consists of three phases: curing, holding, and warming.

Curing. During the first part of storage, hold tubers at a temperature of 50° to 55°F (10.0° to 12.8°C) with relative humidity above 95%. These conditions favor rapid suber-

Figure 24. Pest populations develop resistance to pesticides through genetic selection. Certain individuals in a pest population are less susceptible to a pesticide than other individuals (A). The less-susceptible individuals are more likely to survive the pesticide and produce progeny, some of which are also less susceptible to the pesticide (B). Repeated applications of the pesticide result in a population that is dominated by less-susceptible individuals (C). When this happens, the pesticide and related materials with the same mode of action no longer effectively control the pest.

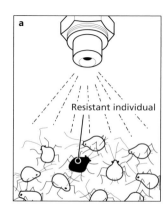
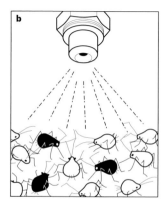
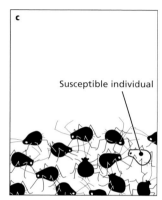

ization of any bruises or cuts incurred during harvest and allow the skin of immature tubers to mature. Both of these processes increase the resistance of tubers to decay. Hold tubers under curing conditions for a minimum of 2 weeks, then lower the temperature by 0.5°F (0.3°C) per day until the desired holding temperature is reached. Do not bring in outside air that is cooler than the pile. If there is increased risk of decay, as with tubers injured by frost or tubers exposed to late blight or excessively wet conditions during harvest, store affected lots separately. Cool them to 50°F (10.0°C) and dry them as quickly as possible with high flows of nonhumidified air. Be careful not to overdry the tubers because loss of tuber weight will occur.

Holding. For most of the storage time, hold tubers at the lowest temperature possible without affecting market quality. The following holding temperatures are recommended:
- chipping: 50° to 55°F (10.0° to 12.8°C); 42° to 45°F (5.6° to 7.2°C) for cultivars resistant to cold sweetening
- french fries: 45° to 50°F (7.2° to 10.0°C); 42° to 44°F (5.6° to 6.7°C) for cultivars resistant to cold sweetening
- fresh market: 40° to 45°F (4.4° to 7.2°C)
- seed: 35° to 40°F (1.7° to 4.4°C)

Maintain high humidity to keep tubers from drying and to avoid pressure bruising but low enough to prevent surface wetness. Some cultivars are more susceptible to pressure bruise; shallower piles may be needed to reduce the likelihood of pressure bruise. Remember that high humidity maintains pile weight, but condensation encourages disease. After proper curing, manage the humidity to prevent surface moisture on tubers. Always visit storages at least once a week to look for storage problems.

Higher temperatures and longer storage times can increase the severity of leafroll net necrosis.

Warming. If holding temperatures were lower than 50°F (10.0°C), letting tubers warm up to this temperature before removing them from storage reduces bruising. Allow the heat of respiration to warm tubers. Do not use warm air; condensation may occur on cold tubers, creating conditions that favor decay. Be sure to maintain humidity to keep tuber water content at the proper level. Tubers with lower water content are more susceptible to blackspot bruising. If excessive sugars have accumulated in tubers to be used for processing, warming above 50°F for 3 weeks may reduce the sugar to acceptable levels. Before cutting seed tubers, warm them at least 10 days at 50° to 55°F (10.0° to 12.8°C) to increase their wound-healing ability. Warmed tubers also cut more easily with less physical damage.

Pesticides

Anyone working with pesticides should be properly licensed, know the identity of the chemicals being used and their potential for injury to humans, livestock, crop plants, or the environment; necessary safety precautions; and what to do in case of an emergency BEFORE USING THEM. ALWAYS READ THE LABEL CAREFULLY BEFORE USING ANY PESTICIDE. Follow directions and observe suggested safety precautions. Be aware of changes in labels of the pesticides you use. Know the target pest for each application and which nontarget species, especially natural enemies and pollinators, may be affected.

Properly used, pesticides can provide economical protection from pests that otherwise would cause significant losses. In many situations, they are the only feasible means of pest control. Careless or excessive use of pesticides, however, can result in poor control, crop damage, higher expenses, and hazards to health and the environment. In an IPM program, pesticides are used only when field monitoring, field history, or disease prediction models indicate they are needed.

In choosing a pesticide, consider not only its effect on the target pest, but also the effects it may have on other pests, natural enemies, and crop plants. Effects may vary according to formulation, rate, and application method. Consult references listed in the suggested reading and local experts for current information on new materials and methods before making a choice.

Make sure each pesticide treatment is suited to the appropriate stage of the target pest and is timed to coincide properly with crop growth, weather conditions, and cultural operations. If you can get the control you need with fewer treatments, you will reduce costs and potential hazards. Also, the chance of such side effects as pest resurgence, secondary pest outbreaks, and pesticide resistance will be lessened.

Pesticide Resistance. Certain pests of potatoes have developed resistance to certain pesticides, that is, they are able to survive applications that once killed most individuals of the same species. These pests include some insects and mites; certain pathogens, such as those that cause late blight, silver scurf, and early blight; and a number of weed species. Table 11 lists pesticides for which resistance has been reported in pests that affect potato. At first, it may be possible to control resistant pests by increasing rates, but higher rates and more frequent exposures increase the chance that resistant individuals will increase as a proportion of the total population (Fig. 24). When that happens, the pesticide involved will no longer be economically useful. The likelihood of resistance can be reduced by rotating materials with different modes of action or using tank mixes of materials with different modes of action against the target pest.

The search for new pesticides is complicated by the fact that species resistant to one material often develop cross-resistance to others, even to materials not in the same chemical class. Resistance to new pesticides may then develop much more quickly than it did with the original one. For example, insect pests that are resistant to chlorinated hydrocarbons such as DDT are more likely to become resistant to pyrethroids. Also, pesticide resistance can develop in a pest not intended as the target of applications.

For certain insects, particularly the Colorado potato beetle, test kits are available that detect resistance to major classes of insecticides. Using these prior to pesticide application can help avoid control failures or exacerbation of resistance if it is already in the population. In areas where resistance is a concern, Colorado potato beetles emerging in the spring can be tested for insecticide resistance before any pesticide application. Colorado potato beetle insecticide resistance test kits are available for this purpose.

Pest Resurgence and Secondary Outbreaks. Application of pesticides that kill or disturb natural enemies may result in pest resurgence or outbreaks of secondary pests. These problems are most common with the use of insecticides.

Pest resurgence occurs when a pesticide destroys both the target pest and its natural enemies. Because the natural enemies depend on the pest for food, they take longer to build up to their former numbers. On the other hand, pests that survive treatment or that move into the field later can breed without the restraint of natural enemies, sometimes increasing to greater numbers than existed before treatment. Systemic insecticides must be used carefully and in a timely manner to minimize impacts on nontarget species.

Pesticide applications may also cause damaging increases in pests that were not targets. Such increases, called secondary outbreaks, usually result from destruction of natural enemies that controlled the secondary pest before the application (Fig. 25). For instance, secondary outbreaks of spider mites may occur in potatoes when natural enemies are destroyed by insecticides applied for aphids or other insect pests. Likewise, application of carbaryl to control chewing insects often results in increased aphid populations. In some areas, extensive use of fungicides to control late blight may also reduce populations of fungi that naturally control aphids within the crop. When using foliar fungicides, diligently monitor aphid populations to prevent this form of secondary outbreak.

The relative toxicity to natural enemies of pesticides registered for use in California can be found in *UC IPM Pest Management Guidelines: Potato* listed in the suggested reading.

Accelerated Breakdown. Breakdown by certain soil microbes is the primary means by which many pesticides, including soil-applied herbicides, are removed from the environment. Repeated applications of certain herbicides, insecticides, or fungicides may encourage populations of these microbes to increase. When this happens, the pesticides are broken down so quickly that they become either less effective or useless.

Table 11. Pests of Potato in the United States for Which Pesticide Resistance Has Been Reported. Resistance in a particular species does not necessarily occur uniformly in all areas. Weed species listed are those for which resistance has been reported in one or more western states.

Pest Species	Pesticide
INSECTS AND MITES	
Leptinotarsa decemlineata (Colorado potato beetle)	carbamates
	chlorinated hydrocarbons
	neonicotinoids
	organophosphates
	pyrethroids
Myzus persicae (green peach aphid)	carbamates
	chlorinated hydrocarbons
	organophosphates
	pyrethroids
Trichoplusia ni (cabbage looper)	chlorinated hydrocarbons
	organophosphates
Tetranychus urticae (twospotted spider mite)	chlorinated hydrocarbons
	organophosphates
PATHOGENS	
Alternaria solani (early blight)	strobilurins
Fusarium spp. (dry rot and seed piece decay)	benzimidazoles
Helminthosporium solani (silver scurf)	benzimidazoles
Phytophthora erythroseptica (pink rot)	phenylamides
Phytophthora infestans (late blight)	phenylamides
Pythium spp. (leak)	phenylamides
WEEDS	
Amaranthus powellii (Powell amaranth)	photosystem II inhibitors[1]
Amaranthus retroflexus (redroot pigweed)	photosystem II inhibitors[1]
Avena fatua (wild oat)	ACCase inhibitors[2]
	ALS inhibitors[3]
	pyrazoliums
	thiocarbamates
Echinocloa crus-galli (barnyardgrass)	ACCase inhibitors[2]
	thiocarbamates
Kochia scoparia (kochia)	ALS inhibitors[3]
	synthetic auxins (growth regulators)
	photosystem II inhibitors[1]
Salsola tragus (Russian thistle)	ALS inhibitors[3]

1. Includes triazines, uracils, phenylcarbamates, and substituted ureas.
2. Includes aryloxyphenoxypropionates and cyclohexanediones.
3. Includes imidazolinones, sulfoanilides, sulfonylureas, and triazolopyrimidines.

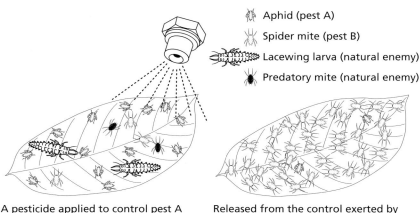

Aphid (pest A)

Spider mite (pest B)

Lacewing larva (natural enemy)

Predatory mite (natural enemy)

A pesticide applied to control pest A also kills natural enemies that are controlling pest B.

Released from the control exerted by natural enemies, pest B builds up to economically damaging levels.

Figure 25. Secondary pest outbreaks are most often induced when pesticides kill natural enemies that have kept a nontarget pest under control. In the absence of natural control, the nontarget pest multiplies to reach damaging levels. Disruption of competition may also cause outbreaks of secondary weed, disease, or insect pests.

In some fields in the Northwest, S-ethyl dipropylthiocarbamate (EPTC) has lost its effectiveness due to the buildup of microbes that degrade it. Avoid repeated applications of EPTC and use the lowest rates necessary for weed control. Problems with accelerated breakdown have also been reported for carbofuran and metalaxyl.

Crop Injury. Crop injury (phytotoxicity) from pesticides can result from an improper application method, poor timing of application, excessive rates, drift, and residues in soil or water. Most common phytotoxicity problems are caused by herbicides. Some potato herbicides may injure certain cultivars; be sure to follow label restrictions regarding cultivars. Metribuzin may injure potato plants if applied within 3 days following cloudy, cool weather. Maleic hydrazide may cause injury if improperly applied.

When following potatoes with a rotation crop, make sure the herbicides you use do not leave residues potentially harmful to that crop. Before planting potatoes, be sure no harmful residues are left over from the rotation crop.

Hazards to Human Health. Some pesticides used in potatoes are hazardous to humans. The applicator is most at risk; field personnel, irrigators, and others who enter the field may also be exposed. Federal law requires safety training for any workers who will enter a field any time within 30 days following the expiration of the restricted-entry interval for a pesticide applied in that field.

Pesticides should not be allowed to drift into roadways, yards, or other areas outside the field where they may cause health problems. Some pesticides may contaminate groundwater supplies. Read and follow the safety precautions on pesticide labels and consult experts for guidelines on how to use pesticides safely and appropriately. Several safety publications are listed in the suggested reading.

Before applying a pesticide, learn what emergency measures can be used safely by field personnel and confirm the availability of emergency health care. Make sure application equipment is in good condition and has the appropriate safety features for the material you are using. Always wear appropriate protective clothing when handling or applying pesticides. Dispose of containers according to label directions.

Hazards to Wildlife. Pesticides applied to potatoes may harm wildlife if runoff or spray drift contaminates bodies of water or nearby natural areas. Nontarget wildlife may be affected by toxic baits used to control vertebrate pests. Stay informed about pesticide regulations that are designed to protect water supplies and wildlife in your area. Take special precautions, select less-harmful materials, or avoid using pesticides where runoff or spray drift is likely to contaminate nearby bodies of water or sensitive wildlife habitat.

In areas where endangered species occur, special guidelines may apply to the types of pesticides used and the means of application. The United States Environmental Protection Agency (EPA) has issued bulletins by county that list the endangered species that occur, map their location, and specify for each pesticide the guidelines that should be followed when using them in the vicinity of endangered species habitat. These county bulletins can be obtained from the EPA World Wide Web site (www.epa.gov/espp) and from county agricultural commissioners. Your county agricultural commissioner will also have information regarding any additional restrictions that may be required by your state.

Organic Potato Production

Potatoes are one of the most difficult crops to grow organically. Production of good-quality organic potatoes is possible, but requires more intensive management than conventional potatoes. Fewer pest control options are available to organic

Table 12. Pesticides Acceptable for Use on Organically Grown Potatoes.

REGISTRATION AND ORGANIC CERTIFICATION OF THESE AND OTHER MATERIALS MAY CHANGE. CONSULT LOCAL AUTHORITIES FOR THE LATEST INFORMATION.

Pesticide	Pest controlled	Comments
azadirachtin	many insect pests	Phytotoxicity; toxic to predators; less effective than Bt formulations.
Bacillus subtilis	early blight, late blight	Late blight is suppressed but not controlled.
Bacillus thuringiensis (Bt)	many caterpillar pests, including loopers and armyworms	Must be applied when larvae are small and feeding; short-residual.
Bacillus thuringiensis subsp. *tenebrionis*	Colorado potato beetle	Must be applied when larvae are small and feeding; short-residual.
Burkholderia cepacia	root diseases caused by *Fusarium*, *Rhizoctonia*, and *Pythium*	
copper-based fungicides	early blight, late blight	Poor control of late blight when used on a 7-day application schedule.
insecticidal soaps	many insect pests	Phytotoxicity.
Pseudomonas syringae strain ESC-10	Fusarium dry rot and seed piece decay, silver scurf	Efficacy unknown.
pyrethrin	aphids, leafhoppers	Highly toxic to predators.
rotenone	many insect pests	Safety considerations; toxic to predators; highly toxic to fish.
spinosad	Colorado potato beetle, foliage-feeding caterpillars, thrips	Some formulations organically acceptable.
sulfur (dust, wettable)	mites, powdery mildew	
sulfur (soil sulfur)	common scab	Effective when soil pH lowered to 5.5.
Trichoderma harzianum strain KRL-AGZ	Fusarium dry rot and seed piece decay, *Rhizoctonia*	Efficacy unknown.

growers. Some chemical and microbial controls are available (Table 12); however, they tend to be less effective than conventional chemical controls. The organic grower must rely more heavily on cultural practices that reduce or avoid pest populations and promote a vigorous, healthy crop that is better able to tolerate pest pressures. Special attention must be given to sanitation practices. Increased monitoring is necessary in many cases to allow susceptible stages of a pest to be targeted, to control pest populations while they are still small, or to remove plants at the first sign of disease. Practices that may reduce pest problems in organic potato fields include the following.

- Use rotations and cover crops to reduce pest populations and increase soil fertility.
- Rotate out of potatoes for at least 2 years and preferably 4 or 5 years.
- Plant as far away as possible from the previous season's potatoes; this can greatly reduce the pressure from pests such as Colorado potato beetle that may have overwintered in last year's potato fields.
- In warmer areas such as the low-elevation desert valleys, avoid growing potatoes during summer months.
- To avoid problems with foliar fungal diseases, select earlier maturing cultivars and cultivars that tend to grow upright with fewer stems, which allows better airflow through the plant canopy.
- Use certified seed potatoes. When organically grown certified seed potatoes are available, their use may be required for certified organic production.
- Use whole seed potatoes, about 2 to 4 ounces (or about 50 to 200 g), to eliminate the risk of infection during cutting operations.
- Avoid planting seed with sprouts long enough that they may be broken; broken sprouts are susceptible to infection by decay organisms.
- Avoid planting when conditions are cold and wet.
- Try to provide sufficient moisture at planting so that no irrigation is needed between planting and emergence.
- Consider using drip irrigation instead of overhead sprinklers to help avoid the wet foliage that favors diseases such as late blight and white mold.
- Provide adequate soil fertility and moisture to favor vigorous plants.
- Use hand-weeding, cultivation, or flaming to control weeds until vines close.
- Consider using floating row covers to exclude aphids, leafhoppers, and Colorado potato beetle, but ONLY in fields where there are no overwintering populations of Colorado potato beetle.

- Using straw mulches may reduce the ability of Colorado potato beetle to find potato plants and may create an environment favorable for predators, especially predaceous ground beetles.
- Keep tubers covered with soil.
- Whenever practical, carefully remove diseased plants as soon as they are seen.
- Cut and remove vines before field-curing tubers, especially when early blight or late blight was present during the season.
- Store only undamaged tubers with no signs of disease.
- In organic seed potato fields, carefully remove diseased plants as soon as symptoms appear; kill and remove vines as soon as possible.

Insects and Related Pests

Insects injure potatoes by transmitting pathogens and by feeding on tubers, seed pieces, leaves, and stems. The green peach aphid, which transmits the virus causing leafroll and tuber net necrosis as well as other potato viruses, is an important pest in many western potato-growing areas. In commercial potato production, aphid monitoring and control are intended to keep net necrosis within acceptable levels in harvested tubers. Management of aphids is especially important in cultivars such as Russet Burbank, which are highly susceptible to net necrosis. Aphid control is critical in seed potatoes, regardless of variety, because of the low tolerance for leafroll and other viral diseases. Although insecticides may be needed in many cases, aphid control is only one aspect of virus management; the use of certified seed, the elimination of culls and volunteer plants, and other cultural practices are also essential.

Pest species that attack tubers or seed pieces occur throughout the region, but their importance varies from one area to another. Wireworms are especially damaging in Idaho, Montana, Oregon, Utah, and Washington. Wireworm monitoring requires keeping records of damage to potatoes and other crops; soil-applied insecticides are occasionally needed for control. In central and southern California, the potato tuberworm is the most important insect pest; a monitoring program based on pheromone traps is available for timing sprays. Other tuber pests, including flea beetles, symphylans, cutworms, false Japanese beetle, and seedcorn maggot, occur either sporadically or only in limited areas. Pests that destroy foliage are most damaging during the midseason period when tubers are developing. The Colorado potato beetle, which feeds on foliage in both the larval and adult stages, is a major pest in Idaho, Oregon, Washington, Wyoming, and portions of Colorado and Montana. Certain systemic insecticides applied against aphids often are effective in controlling the beetle, although foliar insecticide treatments may also be needed, depending on the length of the season and the number of pest generations.

Other foliage pests, such as flea beetles, loopers, cutworms, and spider mites, are widespread but are rarely damaging enough to warrant special control measures. Insecticides applied for green peach aphid or Colorado potato beetle often control these minor pests as well.

The potato psyllid, which damages plants by injecting a toxin during feeding, is a major pest of potatoes in Rocky Mountain states. Psyllid damage occurs sporadically in the low-elevation desert valley areas of California and Arizona. Reports of psyllid damage are increasing in Idaho. Monitoring is the same as for green peach aphid; foliar sprays are needed for control in some seasons.

Biological Control

Some insects found in potato fields are beneficial rather than harmful. Parasitic wasps such as *Hyposoter exiguae* attack loopers, armyworms, and other caterpillars. Parasitic flies known as tachinid flies destroy Colorado potato beetles, and other tachinid fly species attack caterpillars. Predators such as the twospotted stink bug, lady beetles, and damsel bugs are also common in potato fields. Although natural enemies seldom keep major pest insects from reaching damaging levels in potato fields, they can be important in limiting pest populations on other hosts, thus reducing the numbers that invade potato fields. Natural enemies also help to keep minor pests such as loopers and spider mites from becoming more damaging. Use of broad-spectrum insecticides often results in outbreaks of these pests by destroying the natural enemies that kept them in check. Neonicotinoid insecticides have a negative impact on populations of damsel bugs, minute pirate bugs, and some lady beetles.

Predators. General predators such as lady beetles, predatory stink bugs, and lacewings feed on a variety of insects and mites. Predators generally produce several generations a year, and both adults and immature forms usually are predaceous. Each individual predator can often consume a large number of prey.

Lady Beetles. A number of lady beetle species may be found in potatoes in the western states. The most common in most areas of the West are the convergent lady beetle, *Hippodamia convergens*, and the sevenspotted lady beetle, *Coccinella septempunctata*. The pink spotted lady beetle, *Coleomegilla maculata*, occurs commonly east of the Rocky Mountains. Adults are small, dome-shaped beetles about ¼ inch (6 mm) long, with pink, red, or orange wing covers that usually have a pattern of dark spots. Yellow to orange, oblong or spindle-shaped eggs are laid in clusters of less than 20 on leaf surfaces near prey populations. Eggs are similar in appearance to Colorado potato beetle eggs, but smaller; Colorado potato beetle eggs usually are laid in clusters of 30 to 50. Lady beetle larvae are elongated, alligator shaped, and dark blue-gray or black with red, orange, or yellow markings. Both larvae and adults are predaceous, consuming large numbers of prey. They feed primarily on aphids and Colorado potato beetle eggs. Some species also consume potato beetle larvae and eggs of other insect pests. Lady beetles are susceptible to organophosphate and carbamate insecticides and are highly susceptible to pyrethroids.

Green Lacewings. Green lacewings, *Chrysopa* spp. and *Chrysoperla* spp., occur commonly in potato fields. Adult green lacewings are bright green, about ¾ inch (19 mm) long, and slender with four delicate wings that fold over the body like a tent when at rest. Eggs are laid singly or in groups,

Adult lady beetles are small, shiny, and dome shaped. Most are red, orange, or pink, usually with black markings. A sevenspotted lady beetle, *Coccinella septempunctata*, is shown here.

Lady beetle adults and larvae feed on aphids as well as on the eggs of Colorado potato beetle and other insects. The larva of the convergent lady beetle, *Hippodamia convergens*, is shown here.

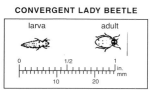

CONVERGENT LADY BEETLE

Lady beetle eggs are similar in appearance to those of Colorado potato beetle, but smaller. Lady beetle eggs usually are not present before mid-season.

depending on species, with each egg attached to a leaf surface by a long, threadlike stalk that helps protect against predation. Larvae are mottled gray or yellow-gray and reach a length of about ⅜ inch (9 mm). They are flat, tapered at each end, and have long curved mandibles with which they grasp and withdraw body fluids from their prey. Lacewing larvae feed on any soft-bodied insects that are small enough for them to capture, including aphids, Colorado potato beetle larvae, small caterpillars, leafhoppers, and psyllids, as well as insect eggs and mites. *Chrysopa* adults are predaceous; *Chrysoperla* adults feed on honeydew and plant nectar.

Stink Bugs. Two species of predaceous stink bug occur commonly in potatoes—the twospotted stink bug, *Perillus bioculatus*, and the spined soldier bug, *Podisus maculiventris*. Adults are shield shaped and about ½ inch (12 mm) long. The twospotted stink bug adult is red or yellow with a black Y on the back. The spined soldier bug is pale brown to tan with

a prominent spur on each shoulder. Stink bugs lay clusters of barrel-shaped eggs on leaves and twigs. Young nymphs of both species are round in shape and red and black. Cream, yellow, and orange markings develop on older nymphs. Nymphs and adults of both species feed on beetle larvae, especially larvae of Colorado potato beetle, and on some caterpillar pests such as cutworms and armyworms. Stink bugs are susceptible to organophosphates and carbamates and are less susceptible to pyrethroids.

Bigeyed Bugs. Bigeyed bugs, *Geocoris* spp., are effective predators of aphids, leafhoppers, and spider mites in potatoes. They are about 3/16 inch (5 mm) long when mature and are light brown, gray, or tan. Their broad head and large eyes help distinguish them from other bugs that may occur in potatoes. Nymphs are light gray or blue and are similar in shape to adults but smaller and without wings. Both nymphs and adults are predaceous, feeding on a variety of insects,

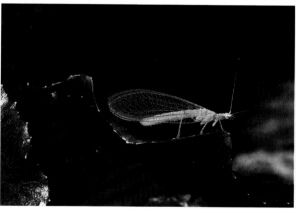

Green lacewing adults are long and slender with delicate wings that are folded over the body when at rest. Adults of some species are predaceous.

GREEN LACEWING

egg   larva   pupa   adult

Green lacewing larvae eat large quantities of mites, aphids, and other small insects and insect eggs. They use their long, curved mandibles to seize prey and withdraw body fluids.

The green lacewing egg is attached to a leaf surface or stem by a long stalk. Some species lay eggs singly, as shown here; others lay eggs in groups.

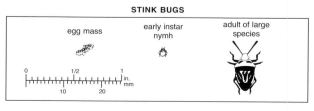

The twospotted stink bug is usually yellow or red, with a black Y on its back and two black spots on its thorax. This one is attacking a Colorado potato beetle larva, but twospsotted stink bugs also feed on cutworms and armyworms.

STINK BUGS

egg mass   early instar nymh   adult of large species

insect eggs, and mites. They also feed on plant nectar, which allows them to survive when prey is not available, but also makes them susceptible to systemic insecticides including neonicotinoids.

Damsel Bugs. Damsel bugs, *Nabis* spp., occur commonly in potatoes. Adults are slender, grayish brown, about ⅜ to ½ inch (9 to 12 mm) long, and have long legs, a narrow head, and bulging eyes. Nymphs are similar in appearance but smaller and lack wings. Both adults and nymphs feed on spider mites, insect eggs, and a wide variety of soft-bodied insects, including aphids, leafhoppers, psyllids, caterpillars, and Colorado potato beetle larvae. Damsel bug populations are reduced by soil applications of neonicotinoid insecticides.

Minute Pirate Bugs. Minute pirate bugs, *Orius* spp., are active predators that feed on all stages of mites, insect eggs, aphids, small caterpillars, and the immature stages of a number of small insects. Adults are about ⅛ inch (3 mm) long, black, with conspicuous silvery triangular markings on the back. Nymphs are smaller and yellow or amber with red eyes. Both nymphs and adults are predaceous. Pirate bug populations are susceptible to soil applications of neonicotinoid insecticides.

Predaceous Ground Beetles. A number of species of predaceous ground beetles, family Carabidae, are found in potato fields. They live on or in the soil where they feed on soil-dwelling insects and the larvae and pupae of insects such as Colorado potato beetle and cutworms that crawl to the ground to

Bigeyed bugs feed on small caterpillars, mites, aphids, leafhoppers, and on the eggs, small nymphs, and young larvae of several other insect pests.

BIGEYED BUG

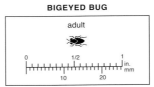

Minute pirate bugs are predators that feed on thrips, psyllids, small aphids, and insect eggs, as well as on the mite eggs shown here.

Minute pirate bug nymphs are yellow or amber with prominent red eyes. They feed on the same prey as adults.

MINUTE PIRATE BUG

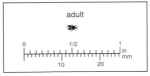

DAMSEL BUG

Damsel bugs feed on different stages of a variety of insects, including aphids, leafhoppers, and psyllids. Both the nymph (right) and adult (left) are predators.

pupate. Some species also feed on aphids. Adults are fast moving, shiny beetles with long legs. Most are black or dark reddish, but some may be brilliantly colored. The larva is elongated and somewhat flattened, with a large head and distinct mandibles. Both larvae and adults are predaceous.

Spiders. A number of spider species may occur in potatoes. They are all general predators, feeding on insects and other arthropods. Dwarf spiders, *Erigone* spp., usually are the most important. They feed on green peach aphid, other aphids, and on spider mites. Ground spiders in the family Gnaphosidae occur commonly, feeding on insects at the soil surface.

Monitoring

As soon as shoots start emerging, begin walking the field regularly to look for blank spots, weak shoots, and infestations on young plants. If you find weak areas or damaged shoots, check seed pieces and look in the soil for soil-dwelling pests such as wireworms, seedcorn maggot, and symphylans. Continue monitoring for pest activity by visiting each field at least once a week throughout the season to make sure the crop is developing normally. The most important monitoring activity, especially where Russet Burbanks are grown, is sampling for aphids; the procedure is described in the green peach aphid section of this chapter. Keep alert for outbreaks of psyllids, spider mites, Colorado potato beetles, and other foliage pests.

Other monitoring activities are needed only in certain situations. In newly cultivated land, in fields planted to potatoes following weedy alfalfa or pasture, and in places where wireworm injury was noted in previous crops, you may need to collect soil samples before planting and sift them to look for wireworms. In parts of California where the potato tuberworm occurs, set out pheromone traps and check them twice a week to monitor activity of the tuberworm adults.

SUCKING INSECTS AND MITES THAT DAMAGE POTATOES

Insects such as aphids and psyllids use their specialized mouthparts to puncture the host plant and withdraw sap from the phloem, or food-conducting tissue. Aphids injure potatoes primarily by spreading *Potato leafroll virus* and *Potato virus Y*. Aphids only occasionally increase to populations high enough that their feeding injures potato plants directly. Psyllid nymphs inject a toxin while feeding that causes psyllid yellows. Adult and immature leafhoppers become a problem when they spread phytoplasmas that cause diseases such as beet leafhopper transmitted virescence and aster yellows (purple top). Spider mites usually are not a problem unless the use of broad-spectrum insecticides destroys populations of mite-feeding predators, causing secondary outbreaks of spider mites.

Predaceous ground beetles are shiny, dark, fast-moving beetles that live on the ground, feeding on soil-dwelling insects and the larvae and pupae of insects that pupate in the soil.

Larvae of predaceous ground beetles are elongated with a large head and prominent mandibles. The body tapers toward the rear, with prominent appendages (cerci) at the tip. They feed on the same prey as adult ground beetles.

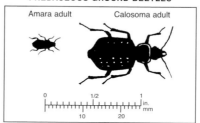

PREDACEOUS GROUND BEETLES

Amara adult　　Calosoma adult

0　　1/2　　1 in.
　　　　　　　　mm
　　10　　20

Dwarf spiders, *Erigone* spp., are very small—no larger than about $1/12$ inch (2 mm). They feed on aphids and spider mites.

Green Peach Aphid
Myzus persicae (Aphididae)

The green peach aphid (GPA) is a pest everywhere potatoes are grown. It is the vector of *Potato leafroll virus* (PLRV), the pathogen causing leafroll and tuber net necrosis, *Potato virus Y* (PVY), and other viruses. Seed potatoes have an especially low tolerance for green peach aphid because the incidence of aphid-transmitted viruses must be very low for seed to be certified. Management involves an integrated program that starts with certified virus-free seed tubers and may include sanitation, the use of sprout inhibitors, weed management, and the use of systemic insecticides early in the season. Monitoring methods are available for making decisions about the need for foliar insecticide applications in some areas. In commercial potato production, green peach aphid generally does not need to be treated on cultivars not susceptible to net necrosis or where growers use seed from meristem culture programs on a wide-scale basis.

Description

Like most aphids, green peach aphid has both winged and wingless forms. Wingless forms are teardrop shaped and usually light green, although some populations, especially in late summer, include many individuals that are pink. In winged forms, the head and thorax are brown or black, and the abdomen is green with a darker patch on the back. Immature aphids that are developing into winged forms have wing pads.

Green peach aphid is by far the most common aphid on potatoes in the western United States, accounting for more than 90% of the aphids found on potatoes in the region. The next most common species is the potato aphid; others are found only occasionally. You can distinguish the green peach aphid from potato and other aphids by size, body shape, and the shape of the antennal tubercles (Fig. 26).

Seasonal Development

The green peach aphid has a complex life cycle with 10 to 25 generations a year (Fig. 27). Most generations consist only of females that produce live nymphs without mating. Sexual reproduction is absent altogether in areas with mild winters (Fig. 28). The population spreads from one host to another all year as new hosts become available and others dry out or are destroyed by frost. The green peach aphid feeds on several hundred plant species, including many crops, weeds, and ornamentals.

In areas with cold winters, a generation of sexual forms appears in the fall and eggs are laid on a winter host—peach, apricot, and certain plums. Overwintering eggs hatch in the spring, producing a generation of wingless females called stem mothers. The stem mothers feed on the buds and young leaves of the winter host; each one produces 100 to 200 wingless females, beginning a series of generations on the winter host.

Starting in the third spring generation, some nymphs develop into winged migrants and fly to other hosts to begin a series of summer generations. When a migrant reaches a new host plant, it usually remains only long enough to produce a few nymphs, then moves to yet another plant. Most flights by migrants are short, so infestations usually spread gradually from the winter host. Migrants can, however, be carried long distances by wind, so even plants far downwind may be infested. Early-spring hosts include mustards and related weeds in and around orchards, but as the season progresses, many plants will become infested.

Each nymph deposited by a spring migrant can start, by asexual reproduction, a new colony of nonmigrant, nonwinged aphids. After one or more generations, the colony

Most green peach aphids are light green; some may be pink. The shape of the tubercles at the base of the antennae distinguish them from other aphids found on potatoes (see Fig. 26).

The winged adult green peach aphid has a dark brown or black head and thorax, and a green abdomen with dark patches. Adults may also be wingless.

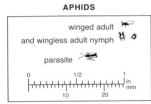

APHIDS

winged adult
and wingless adult nymph

parasite

0   1/2   1
|++++++++++++++++++++++| in.
                          mm
        10   20

Head viewed
from above

Body
outline

Green Peach Aphid

Winged

Tubercles

Wingless

Base of
antenna

Eye

Potato Aphid

Winged

Wingless

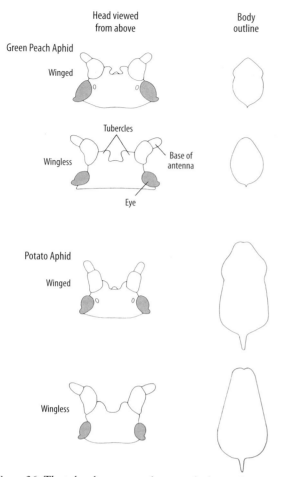

Figure 26. The tubercles converge between the bases of green peach aphid antennae; the body is short and oval in outline. The tubercles of the potato aphid slope outward toward the antennae; the body is elongate.

Figure 28. In the areas of California and Arizona within the gray zone, the green peach aphid develops year-round in a continuous series of generations with no egg stage. In the rest of the western United States, sexual forms appear in the fall and lay eggs on a winter host.

begins to produce a proportion of winged individuals (summer migrants) in each generation. The proportion of migrants to nonwinged adults increases in each generation as colonies become crowded and as host plants dry out or otherwise become less suitable for the aphids. A nymph on a summer host develops to maturity and begins producing a new generation of nymphs in as few as 6 days. In cooler weather, development may take 2 weeks or more.

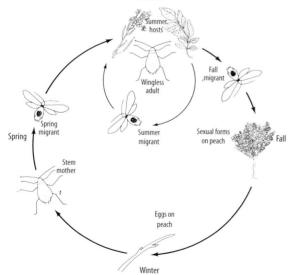

Summer

Summer
hosts

Wingless
adult

Fall
migrant

Summer
migrant

Sexual forms
on peach

Fall

Spring
migrant

Spring

Stem
mother

Eggs on
peach

Winter

Figure 27. Seasonal population cycle of the green peach aphid in areas with a cold winter.

Green peach aphid eggs are laid on or near buds of peach trees or other overwintering hosts. The eggs are dark green at first, then turn shiny black. One black egg can be seen here between the twig and axillary bud.

Winged aphids that colonize potato may come directly from peach trees or other winter hosts, but more commonly they reach the field from intervening hosts such as mustards, nightshades, groundcherries, or ornamentals. Infestations usually advance through the field in a downwind direction. Plants at the edge of the field are infested first and generally remain more heavily infested throughout the season than plants in the center of the field. The aphid population is usually concentrated on the lower third of the plants. Upper leaves may be infested during long periods of cool, cloudy weather or when aphids on lower leaves become crowded.

In most potato-growing areas, the green peach aphid population reaches a peak in the latter half of the season, then declines due to the senescence of host plants, unfavorable temperatures, activity of natural enemies, and emigration of winged forms. Where the season is too short for the population to reach the decline phase, it simply continues increasing through the season.

In late summer or fall, a generation of winged migrants appears that includes both males and females. The females leave the summer host where they developed and fly out at random in search of a winter host. Upon reaching a winter host, they deposit nymphs that become wingless females. Male migrants that reach the winter host mate with these wingless females, which then lay about 5 to 15 overwintering eggs on buds that will open in the spring. Eggs can survive temperatures as low as –50°F (–45.5°C).

The predominant winter host in the western states is peach, *Prunus persica*, although green peach aphid overwinters on other *Prunus* species in some cases. In Idaho, green peach aphid occasionally overwinters on apricot. In Washington, green peach aphid lays overwintering eggs on various *Prunus* species, but winter survival has been demonstrated on a number of weeds such as mallow, filaree (storksbill), and several mustard family weed species. Winter survival on hosts other than peach, including wild *Prunus* species, has been suggested as a possible source of infestations in areas such as the Tule Lake Basin of California and the San Luis Valley of Colorado, where green peach aphid is common even though peach trees are rare. Green peach aphid may survive on American plum, *Prunus americana*, in the San Luis Valley, but not on choke cherry, *Prunus virginiana*. However, choke cherry is a common winter host for green peach aphid in the eastern United States.

The eggs require a certain period of chilling and exposure to water to develop. After the chilling requirement is reached, eggs hatch in response to warm weather. At low elevations, eggs may hatch as early as January or February; at higher elevations or further north, they may not hatch until late April.

In areas with mild winters (see Fig. 28), green peach aphid has no sexual phase; there are no males and no eggs are produced. The population continues moving from one host to another during winter in a series of asexual generations. Even in colder regions, green peach aphid may continue development all winter in greenhouses and other situations where host plants are available and where temperatures remain favorable. Bedding plants, such as vegetables and annual flowers, that are produced in greenhouses and set out in the spring are often infested and can be an important source of green peach aphids that later move to potato. Populations may also overwinter in protected places around heated buildings, along canals, and around springs.

Damage and Relation to Viral Diseases

Injury to potato by green peach aphid is due mostly to transmission of PLRV and PVY. Where a reservoir of the virus exists, a relatively small green peach aphid population can be damaging, especially to seed potatoes. Severe infestations can reduce growth or cause other symptoms by draining plant fluids, but aphids are rarely numerous enough for such injury to be significant in potato production. Aphids are not a major concern in commercial potato production areas where the cultivars grown are not susceptible to leafroll net necrosis.

Transmission of PLRV. To transmit PLRV to a healthy plant, an aphid must first acquire the virus by feeding on a potato plant that is already infected. A period of time called the latent period is required before the aphid can transmit PLRV to a new host plant. The latent period ranges from 10 to 48 hours, but most aphids are capable of transmitting PLRV within 24 hours of acquisition. During the latent period, the virus passes through the aphid's gut, into its blood, and finally into its salivary glands. When the aphid moves to a new host plant and begins feeding, virus particles move with the salivary fluid into the plant. About 30 minutes of feeding is required to produce an infection. Once an aphid acquires the virus, it usually remains infectious for life. The virus does not pass from adult aphids to their nymphs or eggs.

The spread of PLRV to a potato field or from one field to another is due to the movement of winged aphid migrants. Migrants typically live for up to 3 weeks, although most flight activity occurs in the first few days. Wingless aphids can also carry the virus, either by passing directly from leaves of an infected plant to those of adjacent plants, or by crawling over the ground to nearby plants. Both winged and wingless aphids, therefore, can spread PLRV within fields; adjacent plants in the same row as an infected plant are the most vulnerable to virus transmission. Although potato is often the most important source of PLRV, nightshades and groundcherries can serve as reservoirs of the virus. Typically, the green peach aphid transmits PLRV from these weed hosts more efficiently than from potato. Therefore, controlling these weeds can be important in limiting the spread of PLRV.

Transmission of Other Viruses. The green peach aphid is one of several aphids that act as vectors of *Alfalfa mosaic virus* (AMV)(calico) and *Potato virus A* (PVA) and PVY. These viruses, called nonpersistent do not circulate within the aphid's body as does PLRV; they are carried in the aphid vector's foregut and transmission occurs rapidly when contaminated aphids probe host plants. Because the virus particles are quickly lost from the foregut, aphids can transmit them only once or a few times, rather than repeatedly over a long period as in the case of PLRV.

AMV is transmitted mostly by aphids carried by wind from alfalfa fields. Although green peach aphid can carry the virus, aphid species that are more common on alfalfa are the major vectors; these include the pea aphid, *Acyrthosiphon pisum*, and the blue alfalfa aphid, *A. kondoi*. PVY and PVA are generally carried by aphids from infected potato plants, although these viruses can also infect nightshades and other weeds. In addition to green peach aphid, aphid species that do not colonize potatoes but only migrate through the crop can be important in spreading PVY and PVA. Several grain aphids transmit PVY and can be present in large numbers. In commercial potatoes, primary PVY and PVA infections do not reduce yields, although necrotic strains of PVY impact tuber quality. Because of the rapid acquisition and transmission times for these nonpersistent viruses, control of aphids with insecticides is not effective in limiting their spread. In seed potato fields, special cultural control measures may be needed in some cases to limit the spread of nonpersistent viruses by aphids.

Biological Control

Green peach aphids are attacked by a number of predators, including lady beetles, lacewings, damsel bugs, and bigeyed bugs. Several parasitic wasps attack green peach aphids, and the aphids are also susceptible to fungal diseases. Some fungicides destroy populations of the fungi that attack green peach aphid. Although natural enemies cannot be relied on to keep green peach aphids below damaging levels in most situations, their activity helps make other control measures more effective. By choosing insecticides that are least harmful to predators and parasites and using fungicides judiciously, management of green peach aphid is made easier and the likelihood of secondary pest outbreaks and the development of insecticide resistance in aphid populations is reduced.

Management in Commercial Potatoes

The impact of a green peach aphid population in commercial potatoes depends on the cultivar grown and its susceptibility to PLRV and on the number of infected potato plants and weeds that serve as reservoirs of the virus. If the number of reservoir plants is high, even a few aphids can be damaging to a susceptible potato crop. Conversely, if the number of reservoir plants is low, large aphid populations are required

to cause significant spread of the virus. If infected plants are absent, aphids cause little damage unless heavy infestations stress the plants. Since it is seldom possible to be certain that reservoir plants are not present, it is a good strategy to prevent large aphid populations from building up in Russet Burbank and other cultivars susceptible to tuber net necrosis. Cultivar susceptibility to leafroll net necrosis is listed in Table 6 in the previous chapter. In nonsusceptible cultivars or where seed from meristem culture programs keeps virus levels extremely low, treatment of aphids may not be required.

Transmission of PLRV is most likely to result in net necrosis of tubers if it occurs while tubers are bulking, so this is the stage of growth at which aphid control is most important. This growth stage often coincides with the seasonal peak of aphid populations. However, young plants are highly susceptible to PLRV infection, so a few virus-carrying aphids during early growth can significantly increase the incidence of PLRV in the field. Spread by aphid populations that develop later in the season can then result in high amounts of net necrosis in the tubers of Russet Burbank and other susceptible cultivars. Although insecticides are often needed, they are effective only in combination with other practices, including the use of certified seed, the elimination of culls and volunteer plants, roguing of infected plants, weed control, and removal or treatment of aphid overwintering hosts. If there are many infected plants already in the field, no amount of aphid control will limit the spread of PLRV. Therefore, it is important to do everything possible to minimize the number of infected plants before aphids arrive in the field.

Sanitation to Reduce Virus Reservoir. Field sanitation measures such as the elimination of culls and potato volunteers, roguing of infected plants, and weed control, combined with the use of certified seed make up the first line of defense against leafroll. In the absence of these measures, insecticides generally do not prevent even relatively low aphid populations from transmitting damaging levels of PLRV. The widespread use of certified seed from meristem culture programs has greatly reduced the incidence of PLRV. Use of such seed, in combination with good control of volunteers and weed hosts, provides effective control of leafroll in many commercial production areas without the need for insecticides.

The use of infected seed pieces increases the likelihood of leafroll not only in the current year's crop, but also in the next season's, because volunteers may sprout from tubers produced by infected plants and left in the field after harvest. A suitable sprout inhibitor applied to the vines before harvest can reduce emergence of infected volunteers that could serve as a reservoir of virus for the following year's crop. Although it may be more expensive than postharvest treatment, an in-field treatment with inhibitor may be worthwhile in cases where the chance of PLRV infection is high due to a large green peach aphid population or the presence of a large reservoir of PLRV. Follow label directions carefully in timing applications of sprout inhibitor; applying inhibitor too early reduces yield, while applying it too late reduces effectiveness.

Weed control is also an important part of sanitation for reducing green peach aphid populations and PLRV transmission. Follow recommended weed control procedures to keep fields as free as possible of green peach aphid hosts such as mustards, nightshades, and groundcherries. Keep in mind that nightshades and groundcherries can be significant sources of the virus as well as hosts for aphid vectors. If possible, eliminate stands of these weeds in fencerows, ditches, and other uncultivated areas near the field.

Treatment of Winter Hosts. Green peach aphid overwintering in the egg stage (see Fig. 28) is the most vulnerable part of its life cycle. Peach orchards are usually treated in the spring for peach twig borer, leafrollers, or other pests. Some materials used against these pests control overwintering green peach aphid if they are applied after eggs hatch but before winged migrants develop. Oil sprays applied at this time are effective. Also, normal pruning in orchards destroys many green peach aphid eggs. When orchards are abandoned or damaged by frost, however, they may not be pruned or treated for peach pests, and a special treatment for green peach aphid may be worthwhile. The same treatments are recommended for backyard peach trees.

Orchards where the cover crop is a green peach aphid host produce a much larger aphid population than those in which weeds are controlled or the cover crop is a grass. If possible, try to keep peach orchards near potato fields free of weed hosts.

Because untended peach, apricot, and plum trees around homes and volunteer trees along roadsides are often unpruned and surrounded by weeds, these trees may produce exceptionally large numbers of aphids. Removal or treatment of such trees growing near potato fields is highly desirable. Bedding plants can also be sources of green peach aphid. Petunias, green peppers, eggplants, cole crops, and occasionally tomatoes host green peach aphid. Inspect bedding plants for aphids before moving them outdoors; treat them with a suitable insecticide if they are found to be infested.

Community Aphid Control Programs. Because green peach aphid has a large number of hosts and completes much of its life cycle outside potato fields, some practices are effective only on a communitywide basis. In some seed-producing areas this is the only way to prevent virus-carrying aphid populations from moving into potato fields. Model programs established in southeastern Idaho and the San Luis Valley in Colorado have demonstrated that areawide efforts can reduce the incidence of leafroll significantly. The most important practices that require community coordination are the removal or treatment of peach and apricot trees and treatment of greenhouse stock to eliminate green peach aphid from bedding plants such as petunias, green peppers, and eggplants set out in the spring. More background on the Idaho program is available in *Potato Production Systems*, listed in the suggested reading.

It is also important to consider aphid control in early-maturing potato crops prior to vinekill. Winged aphids flying out of these fields often carry PLRV and pose a significant threat to later-maturing cultivars, especially Russet Burbank. Vinekill in early-season cultivars should be as rapid as possible. Slow vinekill, for example by removing irrigation, can further stimulate the production of winged aphids that can carry virus to fields where longer-season cultivars are grown.

Monitoring. Two recommended methods for monitoring green peach aphid are leaf sampling and pan trapping. Leaf sampling is the best method for timing foliar treatments in specific fields. Pan traps are useful for monitoring aphid activity on an areawide basis, but they do not provide a reliable basis for scheduling insecticide applications. By the time winged green peach aphids appear in pan traps, a low level of infestation usually is already present on potato plants.

Leaf Sampling. Methods recommended for leaf sampling vary among the different areas in the western United States, but the following directions are applicable in most cases:

1. Where an orchard, community, or other potential aphid source is nearby, include the side of the field closest to this source in the sample. Start in one corner of the field and move toward the center as you pick sample leaves (Fig. 29). In large fields, move at least 300 to 400 feet (91 to 122 m) into the field before turning back toward the edge. In smaller fields, move to the center before turning back. If potential aphid sources are not nearby, sample all four quadrants of each field (Fig. 30).
2. Pick one leaf from each plant sampled. Remember that potato leaves are compound, each with several leaflets; count aphids on all leaflets of each sampled leaf.
3. Pick sample leaves from the lower third of the plants; this is where green peach aphid is concentrated.
4. Step at least two paces between samples.
5. Depending on the treatment threshold for your area, either count all wingless aphids on the underside of the sample leaf or simply record whether the leaf is infested. If you are counting aphids, you can collect several leaves before stopping to count, because aphids do not readily move off the leaves.

Alternatively, a beating cloth may be used to monitor for aphids. Rather than picking sample leaves, shake the foliage of the lower canopy onto a beating cloth and look for the presence of aphids.

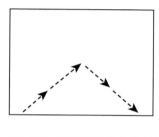

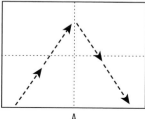

Figure 29. If potential aphid sources are nearby (e.g., an orchard or a community), sample the side of the field closest to the source. Start near one corner of the field and move toward the center as you pick sample leaves.

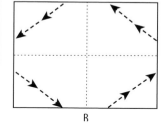

A   B

Figure 30. For a thorough aphid sampling, include all four quadrants of each field by following a pattern such as in A or B. Check field edges in a separate set of samples.

Check potato plants at the edges of the field separately from regular leaf sampling; if an infestation builds up at the edges, it may be worthwhile to apply a strip treatment of foliar insecticide there. In fields where systemic insecticides are applied at planting, there is often an untreated section at the end of the rows due to the space required for the planter to turn around; this section may be infested earlier than the rest of the field and therefore requires special attention in monitoring. Also, watch for green peach aphid on mustards, nightshades, groundcherries, and other weed hosts in and around the field.

Pan Trapping. Pan traps, although not suitable for scheduling treatments, can be used to identify periods when aphids are moving from outside potato fields. A pan trap is a yellow container of water, such as a plastic dishpan, filled to within 3 inches (7.5 cm) of the top; aphids attracted by the color are trapped when they land in the water. A mold inhibitor added to the water helps keep aphid specimens in good condition. Each trap is set on a background of contrasting darker material on the ground, near the crop but not surrounded by vegetation. Pan traps must be checked at least once a week, and often enough so the water does not evaporate between checks.

Several aphid species other than green peach aphid, including potato aphid and some aphids that do not feed on potato, may appear in pan traps. Remove aphids from traps with a forceps or strainer, place them in alcohol, and take them to a lab where they can be examined under a low-power microscope for proper identification. Because of the time required, pan traps are usually operated by local extension agents, farm advisors, growers' cooperatives, or other

agencies that report the results to growers. Pan trapping generally is not worthwhile for the individual grower.

Chemical Control. Insecticides for aphid control in potato include systemic materials as well as foliar sprays. Systemics, applied as seed treatments, at planting, or as sidedressing, control aphids for about 6 to 8 weeks or longer after application, depending on the material. They are important tools where aphids become active early on susceptible crops. Systemic insecticides also control other pests, including Colorado potato beetle and some nematodes. However, seed-treatment and soil-applied systemics are not recommended for aphid control in areas where net necrosis is not a threat.

There is no specific method for deciding whether a systemic insecticide is needed early in the season; make your decisions by evaluating the virus reservoir, the threat of net necrosis, and the potential early-season aphid population based on information from previous seasons. Systemic insecticides prevent aphid populations from building up, but

Pan traps are used to monitor aphid activity on an areawide basis. The yellow color attracts aphids and other insects, which are trapped in the water. These traps can also be used to detect the presence of adult potato psyllids.

they do not kill aphids quickly enough to prevent winged migrants from transmitting viruses. If large numbers of migrants are flying in from virus reservoirs, the incidence of PLRV may be high despite treatment with systemics. If there are infected plants within the field, winged aphids from outside may live long enough to acquire virus from an infected plant and transmit it to a healthy one, even though both plants contain a systemic insecticide. This reemphasizes the need for cooperative, areawide aphid control.

In areas where the virus reservoir is reduced through sanitation and use of certified virus-free seed, aphid populations are less likely to be damaging during the period when systemics are effective. However, because young plants are highly susceptible to virus infection, the presence of systemic insecticide during early development is considered important for avoiding unacceptable levels of net necrosis in most Russet Burbank growing areas. Although systemics are usually applied prior to or at planting, postemergence sidedressing may be recommended instead if slow emergence is expected (for example, when potatoes are planted early in the spring). Otherwise, much of the systemic material may be broken down before plants emerge. Check with local authorities and consult information sources listed in the suggested reading for current recommendations.

Where special risk factors are not present and where aphid populations do not reach damaging levels until the latter half of the season, you can rely on foliar sprays for aphid control if treatment becomes necessary. In these situations, properly timed applications of a suitable insecticide usually provide adequate control. Take into account the results of local aphid survey programs and any information on aphid numbers in other crops to help decide when the level of aphid activity warrants treatment. To schedule foliar treatments, follow the directions for leaf sampling earlier in this chapter and apply insecticide according to current local recommendations. Treatment recommendations for the Pacific Northwest can be found in *Hotline for Aphids on Potatoes* and *Pacific Northwest Insect Management Handbook*, listed under "Insects" in the suggested reading. Recommendations for California can be found in *UC IPM Pest Management Guidelines: Potato* (www.ipm.ucdavis.edu). Check with local authorities for recommendations in other areas.

In some cases, it may be necessary to treat a potato crop to protect other potato fields nearby. For example, a properly timed spray can prevent the movement of winged aphids from varieties such as Shepody, Norland, or Chipeta, which are less susceptible to net necrosis and can tolerate higher green peach aphid populations, to a field of highly susceptible Russet Burbanks. Also, it may be necessary to treat commercial potatoes to prevent aphids from moving to nearby seed potato fields.

Management in Seed Potatoes

Because of the low tolerance for viruses in seed potatoes, more thorough aphid control and extra sanitation practices are needed in seed fields. Essential practices for reducing virus infections include roguing infected plants along with surrounding plants (Fig. 31), harvesting before large aphid populations build up, elimination of volunteer potato plants, selection of seed for increase from the middle of the field, and control of aphids in nearby commercial fields. A single flight of aphids from a field with a high incidence of virus can cause infections in a seed field that will exceed certification tolerances, even if systemic insecticides have been applied. It is easier to keep virus incidence below the tolerance levels required for seed certification in short-season areas where green peach aphid population development is delayed, where host trees are scarce, and where seed fields are isolated from commercial potatoes.

The same control methods are used for green peach aphid in seed as in commercial potatoes, but most areas have no specific treatment thresholds because the tolerance for aphids is so low. The treatment threshold used in Montana is given below. Preplant application of systemic insecticides is recommended in the seed-producing areas of Idaho, Oregon, Washington, and Montana. In some areas, such as the San Luis Valley of Colorado and the Tule Lake Basin of California, it may be possible to wait until aphids appear before applying an aerial spray.

In Montana, if one aphid is found in a field, wait 1 week and sample again. Treat if aphids are still observed. Take a minimum of five samples for 40 acres (16 ha), and one additional sample for each additional 10 acres (4 ha). For each sample take 10 leaves from the lower third of the canopy and 10 leaves from the upper third.

As soon as potato leaves are present, start checking them for green peach aphid. As plants grow, keep checking the lower leaves, where green peach aphids are usually

Figure 31. If you find a plant with leafroll symptoms in a seed potato field, remove and destroy it and adjacent plants in the pattern shown here. Removal of adjacent plants may not be necessary where a systemic insecticide has been used and is still active.

concentrated; aphids are often present even on older, yellowed leaves that are touching the ground. Be sure to check mustards, nightshades, and groundcherries, too; green peach aphid often appears on these weeds before moving to potato. Concentrate monitoring along the edges of the field, especially on the upwind side and in sections closest to peach trees, other green peach aphid hosts, or nearby communities. In some cases, it may be worthwhile to apply a spray only to the edge of the field to destroy an aphid population that might later move into the field.

In fields that have been treated with a systemic insecticide, watch for the appearance of nymphs as a signal that the insecticide is no longer controlling aphids. If you find only winged aphids and a few small, isolated nymphs, the treatment is probably still effective; winged migrants often live long enough in treated fields to deposit a few nymphs. However, if you find wingless aphids surrounded by a group of several nymphs, the aphids must have been feeding for several days and the insecticide is no longer effective.

Results of aphid monitoring programs can be useful in scheduling harvest of seed potatoes; if green peach aphid activity is increasing steadily, you may need to harvest or kill the vines before the number of migrants reaches such a level that the population can no longer be adequately suppressed with sprays.

Potato Aphid
Macrosiphum euphorbiae (Aphididae)

The potato aphid is the only aphid other than green peach aphid commonly found on potato in the western United States. It is generally less damaging because it transmits PLRV much less efficiently than the green peach aphid. However, if the potato aphid reaches very high numbers within a field it may contribute significantly to the spread of PLRV. It also transmits PVY and PVA in a nonpersistent fashion and can be important in the spread of these viruses.

The potato aphid differs morphologically from green peach aphid in the characteristics shown in Figure 26. The potato aphid is larger than the green peach aphid, the body is more elongate, and potato aphids do not have the distinctive converging antennal tubercles found on green peach aphids. Potato aphids may be green or pink. They feed mostly on upper leaves rather than on lower leaves where green peach aphids concentrate.

The life cycle of the potato aphid is similar to that of the green peach aphid. Roses are the only known winter host. Aside from potato, other summer hosts include tomato, groundcherries, and nightshades. Nightshade is a preferred host.

In commercial potatoes, control of potato aphid is needed only in rare cases when extremely large populations stress plants by draining sap. Control may be needed in seed potato fields or in commercial fields in seed-growing areas to limit transmission of viruses. Insecticides for green peach aphid also control potato aphid, so special applications for potato aphid are rarely necessary.

Potato Psyllid
Bactericerca (= Paratrioza) cockerelli (Psyllidae)

Psyllids are small, cicadalike insects about 1/12 inch (2 mm) long. They are related to aphids and leafhoppers and feed in the same way, sucking plant sap through needlelike mouthparts. The potato psyllid occurs throughout the western United States, except parts of the Pacific Northwest, and

Potato aphids are larger and more elongated than green peach aphids. The tubercles at the base of the antennae do not converge as they do in green peach aphids.

Adult potato aphids have winged as well as wingless forms. They may be green or pink.

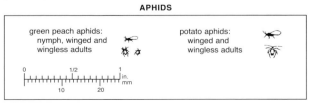

APHIDS

green peach aphids: nymph, winged and wingless adults

potato aphids: winged and wingless adults

0   1/2   1 in.
10   20   mm

in northern Mexico. Damaging infestations on spring- and early-summer-grown potatoes are common in west Texas and New Mexico. Potatoes grown for late summer and fall production are commonly damaged in Colorado, Wyoming, and western Nebraska. Potato psyllid is a migratory insect that winters in the southern United States and Mexico, migrates northward with high temperatures in late spring, and returns south in the fall. Late-spring and early-summer populations are very damaging in northern Mexico and southern U.S. locations. Infestations in other areas arise from these populations.

During feeding, potato psyllid nymphs inject a toxin that causes a disease known as psyllid yellows. Young leaves on affected plants are abnormally erect, their basal portions are cupped, and they often turn red or purple. Affected plants have shortened internodes; older leaves become abnormally

Potato psyllid eggs are borne on stalks at the margins of leaves.

Potato psyllid nymphs have a flattened, scalelike appearance. When feeding, they inject a toxin that causes psyllid yellows.

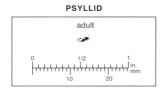
PSYLLID

adult

thick, roll upward, and turn yellow. Plants affected early in the season may be severely stunted or may die. Diseased plants produce a large number of abnormally small tubers that sprout without a dormant period; small aerial tubers may also appear in leaf axils. In fields with a significant number of affected plants, the number of marketable tubers is greatly reduced; tubers from affected plants are not suitable for use as seed. There is a range among potato cultivars in their tolerance to psyllid yellows, but all commercially grown cultivars can be seriously damaged. Psyllid yellows is also an important disease of tomatoes that seriously reduces fruit size and quality in that crop.

As few as 3 to 5 psyllid nymphs per plant early in the season can produce symptoms, but 15 or more are needed to produce severe symptoms. Low populations are damaging before and during tuber initiation, but once tubers are fully formed plants usually tolerate injury. Feeding by adult psyllids has little or no effect at any time.

In the High Plains and Rocky Mountain states, damaging populations occur in most years. Outside the main flight

WILLIAM CALLISON
Psyllid toxin stunts the growth of potato leaves and causes leaflets to roll upward and turn yellow.

Adult potato psyllids are cicadalike in appearance. Although their feeding does not damage potato plants, their presence indicates a need to check for nymphs.

path of the migrating psyllids, outbreaks are less frequent but occur regularly enough to necessitate monitoring every season. Psyllid adults are attracted by the same pan traps used for aphids; sweep nets effectively sample adults but not nymphs. The presence of adults indicates a need to inspect plants for nymphs. Populations in potato fields may be monitored by taking sample leaves in the way described for aphids. Like green peach aphids, psyllid nymphs are concentrated on lower leaves.

Natural enemies of potato psyllid include minute pirate bugs, damsel bugs, bigeyed bugs, the parasitic wasp *Tetrastichus triozae*, and the fungus *Beauveria bassiana*. Outbreaks of psyllids are sometimes associated with the use of systemic carbamate insecticides, which are ineffective against psyllids but reduce populations of natural enemies, such as pirate bugs and damsel bugs, that occasionally feed on plant sap.

There is no specific treatment threshold for potato psyllid. Some insecticides used for aphid control also kill psyllids, but special applications may be needed when psyllids appear earlier in the season than aphids. Neonicotinoids applied at planting, either as seed treatment or in furrow, control psyllids effectively for at least 8 weeks.

Twospotted Spider Mite
Tetranychus urticae (Tetranychidae)

The twospotted spider mite occurs everywhere in the western United States, but significant damage to potatoes occurs only occasionally. The mites injure potato vines by puncturing the surface cells of leaves. Injured leaves develop small yellow splotches that darken to reddish brown and may run together to cover most of the leaf surface; entire leaves may become dried out and brittle. Injury often is not noticed until reddish brown patches of affected plants appear in the field. Injury is most common in hot, dry weather and seldom occurs before midseason. Damage is rare in cooler areas. Severe injury may lower yield by reducing the capacity of plants to perform photosynthesis.

Spider mites are so small that they appear to the naked eye only as tiny, moving dots, although you can see them easily with a 10× hand lens. Adult females, the largest forms, are about 1/60 inch (0.45 mm) long. A single leaf may support many hundreds of individuals. An adult mite has eight legs and an oval body with two red eyespots near the head. Immatures resemble adults and feed in the same way. The name spider mite comes from the loose, thin webbing the mites produce on infested leaves.

Twospotted spider mites feed on a large number of crops and weeds. In cold areas, adults overwinter in soil or in debris on the ground. Infestations in potatoes often begin with adults carried by wind from infested crops such as beans, corn, alfalfa, seed clover, or orchards. Injury often appears first at the upwind edge of the field, then spreads downwind.

The twospotted spider mite has two red eyespots near the head and a dark blotch on each side. The mites are tiny, and a hand lens is required to see these features. Natural enemies, such as the western predatory mite shown here attacking a spider mite, usually keep spider mite populations under control.

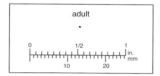

SPIDER MITE

adult

Dust favors the mites; infestations are often concentrated along dusty roads.

Spider mites are often kept under control by natural enemies, including insect predators such as thrips and minute pirate bugs as well as predatory mites. Secondary outbreaks of spider mites may occur when natural enemies are destroyed by insecticides applied for aphids or other insect pests.

When control is needed, a single application of an acaricide is usually sufficient. In some cases, it is necessary only to treat a strip along the edge of the field. Sprinkler irrigation helps limit mite damage by increasing humidity in the plant canopy, making conditions less favorable for the mites.

Leaflets injured by spider mites have yellow blotches and often become dry and brittle. Webbing may be present on heavily infested leaves.

CHEWING INSECTS THAT DAMAGE FOLIAGE

Insects such as beetles and caterpillars can reduce yield if they destroy enough foliage to reduce the ability of the plant to perform photosynthesis. Actively growing plants can tolerate defoliation very well before flowering and tuber initiation. Foliage injury is most damaging during tuber initiation, when a relatively small amount of feeding may reduce the amount of carbohydrate that is available to support tuber initiation and growth. Later in the season, once a dense canopy is developed, plants have more than enough foliage to absorb all the available light. At this stage, plants must lose a significant proportion of their foliage before the damage interferes with light reception and reduces photosynthesis enough to affect yield.

Many chewing insects are controlled by insecticides applied for control of green peach aphid. Special treatments are seldom necessary under western conditions except in areas where the Colorado potato beetle reaches damaging levels.

Colorado Potato Beetle

Leptinotarsa decemlineata (Chrysomelidae)

The Colorado potato beetle occurs in several western states (Fig. 32), and it is an important pest in Idaho, Montana, Oregon, Washington, Wyoming, and northeastern Colorado. It does not occur in Arizona, California, Nevada, or the Klamath Basin in Oregon. Larvae and adults feed on pota-

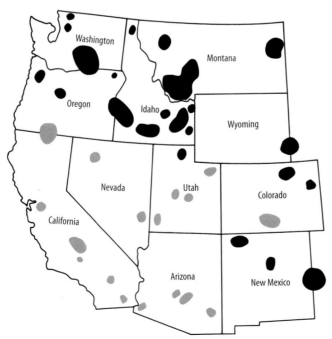

Figure 32. The Colorado potato beetle occurs in the potato-growing areas indicated in black. However, it is not a major pest in New Mexico or Utah.

to foliage; heavy infestations can severely damage vines. Although some insecticides applied for green peach aphid are also effective for potato beetle, extra control measures are needed in many cases.

Description

The oval, convex adult beetle has light yellow wing covers with 10 black stripes; the rest of the body is orange except for black markings on the head and first segment of the thorax. The yellow to orange eggs are laid on end in clusters of 10 to 30, mostly on the undersides of leaves. They resemble eggs of ladybird beetles. Young larvae are dark red with shiny black heads and legs. Older larvae are yellow, orange, or pink, with two rows of black dots along each side. The orange or red pupa is formed in a cell in the soil. The sunflower beetle is similar in appearance. It occurs in growing areas east of the Rockies, where it feeds on wild and cultivated sunflowers but does not attack potato.

Seasonal Development

Depending on the length of the growing season, the Colorado potato beetle has one to three or more generations a year. Adults overwinter in the soil, emerging in spring at about the time volunteer potatoes begin sprouting. Early volunteers are often heavily infested, and nightshades are common early-season hosts. Adults can fly several miles in search of host plants.

Once they find a host plant, the beetles feed, mate, and begin laying eggs. Each female may lay several hundred eggs over a period of a month or more. The eggs hatch in 4 to 10 days, depending on temperature, and the larvae feed for 2 or 3 weeks. Larvae prefer tender young buds and leaves, but feed on increasingly older leaves as they grow. At maturity, each larva burrows into the soil to pupate, forming a pupation cell 4 to 10 inches (10 to 25 cm) below the surface. The pupal stage lasts about a week.

In most areas, adults from first-generation pupae emerge in a few days and lay eggs for a second generation. In the coolest areas, however, there is only one generation a year; these adults emerge from the pupation cell, feed for 7 to 10 days, then renter the soil, where they overwinter. In the warmest areas, there are three or more generations and the overwintering period is much shorter.

The potato beetle was first found in the mid-nineteenth century in the area of what is now the Iowa-Nebraska border, before potatoes were cultivated there. It fed on buffalobur, *Solanum rostratum*, a native plant in the potato family. Most potato family weeds, including black nightshade, cutleaf nightshade, hairy nightshade, and groundcherries, are also potato beetle hosts. Black nightshade is attractive to egg-laying females, but is toxic to the larvae. Weeds and native host plants serve as a reservoir of Colorado potato beetle; there is always at least a small population on these hosts that will

Colorado potato beetles are orange with black-striped yellow wing covers and black markings on the head. The female lays clusters of yellow to orange eggs, usually on the undersides of leaflets. Lady beetle eggs are easily confused with Colorado potato beetle eggs; however, potato beetle eggs are larger and adults usually remain nearby.

The newly hatched Colorado potato beetle larva is dark red with a shiny black head and legs.

Colorado potato beetle larvae and adults chew irregularly shaped holes within and along leaf margins. When populations are high, they can completely defoliate plants.

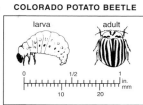

COLORADO POTATO BEETLE

larva   adult

0   1/2   1 in.
                      mm
         10   20

Older Colorado potato beetle larvae are yellow, orange, or pink, and have two rows of black dots along each side.

move to potato plants when they are available. Tomato, pepper, eggplant, and tobacco are potato beetle hosts. Tomatoes can be important alternate hosts in Utah; otherwise no cultivated plant is an important part of the reservoir in the western United States.

Damage

Most potato beetle injury is caused by the larvae, although adults also feed on foliage and buds. Loss of leaf surface reduces the ability of plants to produce carbohydrates for storage in tubers. Heavily defoliated plants have reduced yield and lower tuber quality. The degree of loss depends on the extent of defoliation and the time of season at which it occurs. The greatest losses occur when there is prolonged, extensive defoliation during the period when tubers are

forming. Before and after tuber formation, plants can tolerate moderate defoliation with little or no loss of yield.

Management

Control of Colorado potato beetle is needed most often in areas with two or more generations. In short-season areas, winter mortality often prevents the single generation from reaching damaging numbers. Also, certain systemic insecticides applied for control of green peach aphid frequently control the first generation of potato beetle.

There are no specific treatment thresholds for Colorado potato beetle. When an insecticide is needed, the timing depends on whether you are treating the first or second generation. If you are treating the first generation in an area with two or more generations, spray as soon as you find the first mature (fourth-instar) larvae. If you wait too long, some larvae will complete feeding and enter the soil to pupate, escaping the treatment and emerging later to begin the second generation. If you treat too early, when larvae are in the first instar, overwintering adults may emerge after the spray residue has lost effectiveness. If you are treating the second generation, spray when all the eggs have hatched but while most larvae are still small, before they cause substantial defoliation. It is important to wait until all eggs have hatched; some defoliation at this stage is not harmful. Using a material with relatively long residual action reduces the chance that a second treatment will be needed later.

Colorado potato beetle has shown substantial resistance to major insecticides in the eastern United States (see Table 11), although resistance has not yet become a problem in the West. To reduce the chance that resistance will develop here, limit your insecticide treatments to situations where they are necessary to prevent economic loss and rotate the types of chemicals you use. In areas where they are available, use Colorado potato beetle insecticide resistance test kits to determine if resistance is present in the local population and to help decide which type of insecticide to apply. It is best to use these test kits on spring-emerging beetles early in the season. If the emerging beetle population is too small to cause significant damage, this will allow you to delay treatment until first-generation larvae have developed, by which time you will have the test results. Test results on spring-emerging adults are applicable throughout the season because subsequent generations are derived from the emerging adults.

Elimination of culls and volunteer potato plants, as required for control of viruses and other diseases, helps limit early potato beetle populations. Tillage and crop rotation reduce the number of overwintering adults emerging and surviving in the spring. Rotation to nonhosts and controlling potato beetle weed hosts reduces survival because adult beetles do not travel long distances in search of hosts.

Natural enemies of the potato beetle include lady beetles, which feed on the eggs and young larvae, and predatory stink bugs, which attack eggs and larvae. Lacewing larvae and predaceous ground beetles also feed on potato beetle larvae. Tachinid flies and hymenopteran wasps are common parasites in some areas. Natural enemies do have a significant impact on beetle populations, and in some areas they keep the first generation of larvae from reaching damaging levels. Application of more selective materials such as *Bacillus thuringiensis tenebrionis* formulations and spinosad help preserve this natural control.

Cutworms
(Noctuidae)

Cutworms are caterpillars that are active mainly at night. During the day, they usually hide in soil cracks, under clods, or in debris on the ground. In a potato field with a good plant canopy, they hide under the plants on the soil surface. Cutworms feed mostly on foliage, but they may also chew shallow holes in exposed tubers or cut stems of small plants near the soil surface.

The most common cutworm species that damage potatoes in the western United States are the black cutworm (*Agrotis ipsilon*), the variegated cutworm (*Peridroma saucia*), the spotted cutworm (*Amathes c-nigrum*), the army cutworm (*Euxoa auxiliaris*), and the red-backed cutworm (*E. ochrogaster*). Most cutworms are about 1 to 2 inches (2.5 to 5 cm) long when mature; they vary in color, but generally appear smooth-skinned and often have a wet or greasy appearance. Many species curl into a C when disturbed.

Cutworm adults, drab gray or brown moths, lay their eggs on stems or on the soil surface. Larvae of most species feed mainly on plant parts in or near the soil, although some species may climb onto potato vines at night to feed on higher foliage. Pupation occurs in a cell in the soil. Cutworms overwinter in the soil either as larvae or pupae. There may be from one to three generations a year, depending on the species, climate, and the length of the season.

Cutworm damage is sometimes blamed on loopers or armyworms, which are usually present on plants in the daytime if their damage is noticeable. If you see plants with ragged holes in the leaves but you cannot find caterpillars, grasshoppers, or other chewing pests on foliage, check for cutworms in the soil or on the soil surface under the plants. Damage to tubers and stems is spotty.

Insecticides applied to foliage for control of green peach aphid, potato beetle, and other pests usually kill cutworms too. Cutworms are commonly controlled by parasitic wasps and tachinid flies, predators, and diseases.

Most cutworms, such as the variegated cutworm shown here, appear smooth-skinned and curl into a C shape when disturbed.

Loopers

Trichoplusia ni and *Autographa californica* (Noctuidae)

Loopers feed mostly on older leaves, where mature larvae may chew large, ragged holes. Damage usually occurs late in the season and seldom affects yield even when large numbers of loopers are present. Cutworm injury is sometimes confused with looper injury since the two species may be present at the same time, but loopers are readily found on plants in the daytime while cutworms avoid sunlight, hiding under plants, debris, or in the soil.

The two looper species found on potato, cabbage looper, *Trichoplusia ni*, and alfalfa looper, *Autographa californica*, are very similar. Both are usually light green with several narrow, pale stripes along the back and sides. A small proportion of alfalfa loopers are gray instead of green, with a black head and dark stripes. Loopers are easily recognized by their habit of arching into a loop as they crawl. They have only two pairs of prolegs in the middle of the body, whereas most other caterpillars have four pairs.

In most western potato-growing areas, loopers have two or three generations a year. They feed on a variety of weeds and on alfalfa, beans, vegetables, and cotton as well as on potato. Loopers overwinter as pupae in the soil.

Natural enemies normally keep loopers from reaching damaging levels. In some areas, loopers appear only following the use of systemic insecticides, which kill some natural enemies. Looper eggs and small larvae are attacked by bigeyed bugs, minute pirate bugs, lacewings, and other predators. *Trichogramma* parasites kill the eggs, and several other parasites, especially *Hyposoter exiguae* and *Copidosoma truncatellum*, attack the larvae. Loopers are also subject to a nuclear polyhedrosis virus that can reduce populations rapidly. Common insecticides applied to foliage for green peach aphid and Colorado potato beetle also control loopers. Formulations of *Bacillus thuringiensis* control smaller looper larvae.

Loopers usually are light green with narrow, pale stripes along the back and sides. They arch into a loop as they crawl. Unlike cutworms, loopers are found on potato plants during the day.

Loopers chew large holes, mostly in mature leaves.

Outbreaks of a nuclear polyhedrosis virus can reduce looper populations rapidly. The limp bodies of loopers killed by the virus are left hanging from foliage, often oozing their contents.

CABBAGE LOOPER

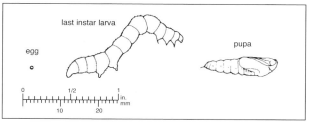

last instar larva

egg

pupa

Yellowstriped Armyworms
Spodoptera ornithogalli and *S. praefica*
(Noctuidae)

These two yellowstriped armyworm species occasionally migrate to potato from infested alfalfa, usually just after the alfalfa is cut. When numerous, the larvae can defoliate plants at the edge of a field, and they may damage exposed tubers in the same way as cutworms. Physical barriers can stop migrating larvae. Plow a trench with the steep side toward the potato field, then apply an insecticide to kill larvae in the trench. Large-diameter irrigation pipe or a strip of stiff aluminum foil set on edge in the soil can also act as a barrier. Insecticide baits are also available. Strip or spot treatments may be needed at the edge of a field, but it is rarely necessary to treat a whole potato field for yellowstriped armyworms. Occasional fieldwide outbreaks of armyworms in California have been associated with the use of systemic carbamate insecticides. Smaller armyworms can be controlled with formulations of *Bacillus thuringiensis*.

The two species of yellowstriped armyworm are too similar to distinguish easily in the field; both have contrasting black and yellow stripes along the entire length of the body, and both have a velvety black spot on the side of the first abdominal segment. *Spodoptera praefica*, known as the western yellowstriped armyworm, is more common in most potato-growing areas of the western states; *S. ornithogalli* is found in southern California, Arizona, and Utah.

Grasshoppers (Acrididae)

Normally, grasshoppers are of little concern in cultivated crops, but in seasons when conditions favor them, large populations may build up on rangelands and may then migrate to fields in damaging numbers. During such outbreaks, grasshoppers can defoliate potatoes and other crops. Hazard maps for the western U.S. are developed each year by the U.S. Department of Agriculture's Agricultural Research Service and the Plant Pest Quarantine division of the Animal and Plant Health Inspection Service. These maps can be used to predict grasshopper populations.

Many grasshopper species that occur in the western United States, but only five or six are crop pests. Common species that affect potatoes in the West include the migratory grasshopper (*Melanoplus sanguinipes*), the redlegged grasshopper (*M. femurrubrum*), the differential grasshopper (*M. differentialis*), the twostriped grasshopper (*M. bivittatus*), and the clearwinged grasshopper (*Camnula pellucida*). Other species are found locally.

Damage is usually limited to a short period in summer, since there is only one generation a year. When grasshoppers move into a field from adjacent land, you can usually control them by spraying the edge of the field or the fencerows or roadsides closest to the source of migration. Insecticidal baits can be effective when applied on an areawide basis to rangelands before migration occurs, when grasshoppers are in immature stages.

PESTS THAT DAMAGE TUBERS

With the exception of the potato tuberworm, insects and related pests that feed directly on tubers are soil-dwelling species; tubers may be injured with little or no aboveground sign of infestation. Infestations are usually spotty, but they often recur from year to year in the same places. Keep records of injury so you can plan preventive measures for subsequent crops.

The western yellowstriped armyworm has black and yellow stripes along the entire length of the body, and a velvety black spot on the side of the first abdominal segment.

YELLOWSTRIPED ARMYWORM

last instar larva

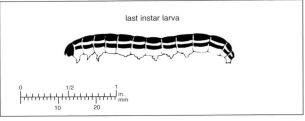

The parasitic wasp *Hyposoter exiguae* lays its egg in a young armyworm larva. This wasp is an important natural enemy of armyworms and loopers.

HYPOSOTER

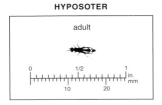

adult

Potato Tuberworm
Phthorimaea operculella (Gelechiidae)

The potato tuberworm is a worldwide pest of stored potatoes. In the western United States, field infestations occur regularly in California from the Sacramento–San Joaquin Delta southward, where warmer winters allow overwinter survival every year. It has recently become a problem in the Columbia Basin and has been found in Arizona, New Mexico, and southwestern Utah. Outdoor infestations elsewhere in the western region usually originate in infested tuber storage. Cultural practices can greatly reduce field infestations; insecticide treatments can be scheduled with pheromone traps.

Description

The tuberworm is a small caterpillar, the larva of an inconspicuous gray and black moth called the potato tubermoth. A larva is usually dull white with a brown head; a mature larva may have a pink or greenish tinge. The smooth, brown pupa is formed in a silk cocoon covered with soil particles or bits of plant debris. The spherical, whitish to brown eggs are seldom noticed because of their small size.

Seasonal Development

The tuberworm has several generations a year in California; development continues all year as long as host material is available. In addition to potato, tuberworms feed on plants in the same family as potato, including tomato, eggplant, black nightshade, silverleaf nightshade, and jimsonweed. Most infestations, however, originate in infested potato plants or tubers in potato fields and cull piles.

Adult female moths lay eggs on foliage, soil, plant debris, or exposed tubers; they will crawl through soil cracks or burrow a short distance through loose soil to reach tubers. On tubers, young larvae feed just below the skin at first, but eventually tunnel deep into the flesh. A newly hatched larva on foliage usually begins feeding between the surfaces of a leaf, creating a small hollowed-out blotch; later, larvae sometimes fold sections of leaf into shelters fastened together with silk. Larvae may also bore into petioles or stems. Generally, however, there is relatively little foliage feeding by tuberworms. Larvae often move from foliage to tubers, leaving the first feeding site, dropping down to the soil, and crawling through soil to tubers; they do not reach tubers by boring through stems or roots.

When they have finished feeding, larvae spin silk cocoons on the soil surface or in debris under the plant; pupation normally does not occur in tubers. In storage, larvae may crawl a considerable distance before pupating in crevices among building materials, in potato sacks, or at a similarly protected site.

Soon after they emerge from pupae, adults mate and females begin laying eggs. Females release a chemical scent, or pheromone, that attracts males for mating. Adults are active at night and at dusk. During the day, they hide in sheltered parts of the plant or on the ground.

The tuberworm completes a generation in 3 to 4 weeks in summer; about half of this time is spent in the larval stage. In winter, larvae may require 3 months or more to complete development. The developmental threshold is about 52°F (11°C); about 700 degree-days are required for a generation. Larvae and adults can survive long periods at temperatures near freezing; feeding and reproduction resume whenever the temperature again reaches 52°F.

The inconspicuous, gray tubermoth is active at dusk or during the night. It hides during the day in sheltered parts of the plant or on the ground.

The potato tuberworm is a small, light-colored caterpillar with a brown head. Newly hatched tuberworm larvae create small hollowed-out blotches in the surface of potato leaves.

POTATO TUBERWORM

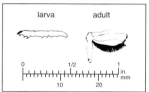

Leaflets damaged by tuberworms are usually curled and shriveled.

Damage

Tuberworms create deep tunnels about ⅛ inch (3 mm) in diameter through the flesh of tubers. The tunnels usually appear to be black, because they are filled with larval feces and are often infected with fungi. Infested tubers are unmarketable. Injury to foliage is rarely extensive enough under California conditions to be damaging.

Management

Anything that reduces exposure of tubers to egg-laying females reduces tuberworm damage. The moths generally cannot reach tubers covered with 2 inches (5 cm) or more of soil unless the soil has deep cracks.

Potato cultivars that set tubers on relatively deep stolons are less vulnerable to infestation; shallow-setting cultivars are more vulnerable. A low level of injury is more damaging in tubers intended for the fresh market than in those intended for processing.

Sprinkler irrigation is valuable in keeping the soil surface sealed, especially in fine-textured soils that tend to develop deep cracks as they dry. Use sprinklers to keep the upper layer of soil moist enough so it will not crack. Tuberworm damage is almost always more severe in furrow-irrigated fields. Where it is necessary to use furrow rather than sprinkler irrigation, make sure the field is properly graded and keep furrows in good shape to prevent washouts that expose tubers.

Prompt, thorough harvest and sanitation are also essential. Harvest tubers as soon as possible after they have matured; do not leave them in the ground longer than necessary. During harvest, avoid leaving tubers on the surface overnight. After harvest, make sure that all unharvested or discarded tubers are deeply buried or destroyed; never leave cull piles in or near the field or around processing or storage facilities. Eliminate volunteer potato plants from fields, waste areas, and from stands of other crops following potato.

Monitoring and Control. Commercial pheromone lures for tubermoths can be used to monitor the pest either in water pan traps or delta traps. Water pan traps consist of a pan about 8 inches (20 cm) in diameter filled with water to a depth of at least 3 inches (7.5 cm); the lure is placed beneath a piece of sheet metal that is bent to form a tent-shaped cover over the pan. Male moths attracted to the lure fall into the water. The water must contain a few drops of soap to break the surface tension and prevent moths from escaping. A delta trap is a three-sided cardboard shelter with a sticky inner surface to catch the moths. The pheromone lure is usually placed on the middle of the sticky surface. Delta traps must be fastened to a stake set in the ground.

Pan traps are generally best for tuberworm monitoring, as they are easily cleaned and refilled at each check, and they are not affected by sprinkler irrigation. The sticky surface of delta traps quickly becomes clogged with dust and moth scales, so the traps must be replaced frequently; when they are dusty, sticky delta traps do not retain the moths attracted to the lure.

Treatment Thresholds Using Water Pan Traps. Use at least four pan traps per field, placing one or more in each quarter of the field. Put the traps between plants on the tops of beds, and keep them at least 14 rows or 50 feet (15 m) in from the edge of the field. Check the traps every 3 or 4 days to count the trapped moths and replenish water in the traps. In hot weather, be sure to check often enough to keep the water from evaporating between checks. For each field, record the number of moths per trap per night (MTN) and keep the results in a chart (Fig. 33).

The treatment threshold varies according to potato variety and field conditions. Fields where tubers are well protect-

Tuberworm larvae feed just below the skin of the tuber at first but eventually tunnel deep into the flesh. Tunnels turn black with larval feces and fungal growth.

Water pan traps for monitoring tuberworm populations use pheromone lures to attract adult males.

ed by soil can tolerate more moths than fields where many tubers are exposed. In fresh market potatoes, treat whenever the MTN for a single check period reaches 15 to 20 *or* when the average MTN for the season reaches 10.

If moth activity does not reach the threshold level before vinekill, do not treat. Insecticides applied at vinekill do not reduce tuberworm damage.

In addition to pheromone trapping, take a random sample of harvested tubers from each field to determine the propor-

Date	MTN this check	Cumulative MTN	Number of check dates	Seasonal average
MAR 1	2	2	1	2
4	5	7	2	3.5
8	3	10	3	3.3
11	6.2	16.2	4	4.1
15	7	23.2	5	4.6
18	9.4	32.6	6	5.4
22	8	40.6	7	5.8
25	10.6	51.2	8	6.4
29	10.8	62.0	9	6.9
APR 2	14	76.0	10	7.6
5	16	92.0	11	8.4

Figure 33. When monitoring potato tuberworm, keep a chart such as this for each field to record the results of pheromone trapping. On each check date, record the number of moths per trap per night (MTN) caught since the previous check. Divide the cumulative total of the MTN values for the season to date by the number of check dates to find the seasonal average MTN.

tion damaged by tuberworm. Check at least 100 tubers taken at random from each quarter of the field. Pay close attention to green tubers; they were exposed and will show more damage. Keep this information as a guide to treatment thresholds the following year: if there is too much damage, you may need to treat at a lower MTN level next year or take extra precautions to protect tubers from exposure to the moths.

Wireworms (Elateridae)

Wireworms, the soil-dwelling larvae of click beetles, bore into seed pieces and tubers, causing internal and external damage. They occur throughout the western United States, but significant damage is uncommon except in Idaho, Oregon, Washington, and Utah.

Description

Several wireworm species occur in western potato soils, but the most common are the sugarbeet wireworm (*Limonius californicus*), the Pacific Coast wireworm (*L. canus*), and the Great Basin wireworm (*Ctenicera pruinina*). All are similar in appearance. Wireworms resemble the mealworms sold as pet food; they are slender, elongate, yellowish to brown larvae with smooth, tough skin. The body is nearly cylindrical, but flat on the lower side. The six short legs are close together near the head, and the tip of the abdomen bears a flattened plate with a pair of short hooks. The slender, brown to black adults are known as click beetles because of their ability to flip themselves over with an audible click when placed on their backs.

Seasonal Development

Common wireworm species require 3 to 4 years to complete their life cycle. Most of the time is spent in the larval stage, but all stages may be present at once during the growing season. Larvae feed on the roots of many crops and weeds and bore into stems and other plant structures such as potato tubers. The larvae move up and down in the soil in response to temperature and moisture. In potato fields, they generally spend most of the growing season in the upper few inches of soil. If the temperature in shallow soil exceeds about 80°F (26°C), however, they move deeper. In areas with cold winters, they overwinter at depths as great as 2 feet (0.6 m), then return to surface layers when the soil temperature reaches about 50°F (10°C). Most damage to potato is caused by larvae in the second and third years of development.

Mature larvae form a pupation cell of soil particles; they may pupate right away or may remain in the cell over the winter, pupating in the spring. Adults that develop in the fall may also remain in pupation cells over the winter. In spring and summer, adults burrow to the surface. Both sexes are capable of flying to reach mates and egg-laying sites. Females burrow back into the soil to lay eggs.

Wireworms, such as the sugarbeet wireworm shown here, are slender yellowish to brown larvae with 6 short legs that are close together near the head.

The tip of the wireworm abdomen is flattened and has a pair of short hooks. The sugarbeet wireworm is shown here.

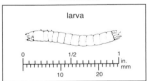

The Pacific Coast and sugarbeet wireworms are found mostly in irrigated soils. Large numbers often build up in fields planted to cereals. The Great Basin wireworm tolerates drier, colder soil conditions than the others; it is the most damaging species in newly cultivated desert soils and in fields recently converted from dryland farming or pasture to row crops such as potato. Infestations of Great Basin wireworms decline over 2 to 4 years of irrigation as the initial population matures and emerges as adults. They may be replaced, however, by sugarbeet wireworms or other species that favor irrigated soils.

Damage

In spring, wireworms may bore into seed pieces and developing shoots. These injuries often become infected with fungi or other decay organisms, resulting in weak plants and spotty stands. In summer, wireworms bore into developing tubers. Entry holes are round, usually about ⅛ inch (3 mm) in diameter. Damaged tubers often look like they have been punctured with a nail. Tunnels in the tubers are usually straight and may be shallow or quite deep. They are usually lined with periderm and are seldom infected with fungi.

Management

Because wireworms have a long life cycle and feed on a variety of hosts, knowledge of the cropping history in a particular field is essential. Wireworms are often present following cereals, even though injury was not apparent in the grain crop, and they may also be present in newly cultivated soils. Although alfalfa is not a favorable host for wireworms, populations often build up in alfalfa fields that are allowed to become weedy.

Where row crops such as potatoes, corn, or beans are planted for several years, crop injury may recur each season in places where wireworms are concentrated. In these situations, it is a good idea to check for wireworms before planting potatoes.

If the soil is not too wet or too fine in texture, you can sample for wireworms with a set of two screens, the upper made of ¼-inch (6-mm) hardware cloth, and the lower of window screen. Collect samples of soil at random in as many places as possible throughout the field, sift them through the screens, and record the number of wireworms. Treatment thresholds are expressed in terms of the number of wireworms per square foot, so you must know the area covered by each sample. For example, a typical two-handled posthole digger removes about ¼ square foot of soil. If you are sampling in

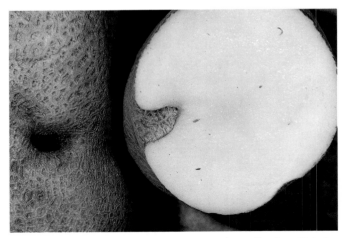

Tunnels made in tubers by wireworms are round, usually about ⅛ inch (3 mm) in diameter, and are usually lined with skin (periderm).

the spring, make sure the soil temperature is at least 45°F (7.2°C) and take samples to a depth of 6 inches (15 cm). In late summer, when surface soil is dry, sample to a depth of 18 inches (45 cm).

You can also check for wireworms by baiting. Bury pieces of carrot about 3 inches (7.5 cm) deep at a series of random locations in the field; mark each place clearly. Check the bait in 2 or 3 days to see if wireworms are present. Another suitable bait is 2 to 3 tablespoons of coarse whole-wheat flour wrapped in a small square of netting. Baiting is not effective if the soil contains a large amount of plant residue, or if the soil is too cold, wet, or dry for wireworms to be active near the surface. If wireworms are detected with baiting, use soil sampling to estimate the population density.

Insecticides for wireworm control may be applied before planting as either broadcast or banded treatments.

Treatment thresholds for preplant applications are usually based on the number of wireworms in soil samples. There are no thresholds based on baiting. Treatment thresholds for California based on soil sampling are given in *UC IPM Pest Management Guidelines: Potato* (www.ipm.ucdavis.edu). Check with local authorities for recommendations in other areas. More information on wireworm monitoring and control is presented in *Potato Production Systems*, listed in the suggested reading.

Once wireworm populations have been reduced, they usually remain low unless favorable crops such as grains are replanted. Where wireworms have been a serious problem, it may be best to avoid grains in rotations with potato. Alfalfa is a good rotation crop for infested soils as long as the stand is strong and does not become weedy.

Flea Beetles (Chrysomelidae)

Epitrix spp.

The tuber flea beetle, *Epitrix tuberis*, is found in the Columbia Basin and Rocky Mountain growing areas. Adults are about 1/16 inch (1.5 mm) long, shiny green to brown or black, with enlarged hind legs that enable them to jump several inches when disturbed. The larvae are whitish, soft-bodied grubs with a yellowish or light brown head and six short legs; they are up to 1/4 inch (6 mm) long.

Adult flea beetles overwinter in weeds or debris outside the field. In spring, they feed on weeds until potato plants emerge, then fly into potato fields and feed on foliage. The adults chew small, round holes in leaves, giving them a shot-hole appearance; this injury is seldom important, although heavy infestations of tuber flea beetle can cause significant defoliation of young plants.

Most damage is caused by the larvae, which hatch from eggs scattered by adult females in the soil around potato plants. The larvae feed on roots, underground stems, and tubers. Larvae pupate in the soil. Development takes about a month and there are one or two generations a year.

Larvae feeding on tubers produce narrow, straight tunnels about 1/32 inch (0.8 mm) in diameter in the flesh of tubers. Fungi often grow in the tunnels, turning them black or brown. Tunnels of the tuber flea beetle may extend 1/2 inch (12 mm) into the tuber; when extensive, they make tubers unsuitable for processing.

Systemic insecticides and foliar sprays applied for green peach aphid and Colorado potato beetle usually keep flea beetles below economically damaging levels. Even in areas where these treatments are not used, flea beetle infestations are sporadic and special controls are rarely necessary. Flea beetle damage often is not noticed until harvest, when it is too late for control measures.

White Grubs (Scarabaeidae)

White grubs are the larvae of scarab beetles. They are widespread in the western states, but they do not commonly cause significant damage to potatoes. Species that feed on potatoes include the carrot beetle (*Bothynus gibbosus*), the tenlined June beetle (*Polyphylla decemlineata*), and the false Japanese beetle, also called sandhill chafer (*Strigoderma arbicola*).

White grubs are usually curled into a C shape. Most of the body consists of the whitish, translucent abdomen, with a dull gray area at the tip. The head and the prominent legs are brown. Larvae may be up to 1 1/4 inches (3 cm) long when mature. The carrot beetle and false Japanese beetle have one generation a year; the tenlined June beetle requires 2 to 3 years to develop. Most of this time is spent as larvae. All species pupate in the soil; adults emerge to mate and lay eggs.

White grubs chew large, shallow gouges in tubers. In some cases, these injuries may cover much of the surface.

EUGENE MEMMLER

FLEA BEETLE

larva   adult

0   1/2   1
|⊢⊢⊢⊢⊢⊢⊢⊢⊢⊢⊢⊢⊢⊢⊢⊢⊢⊢| in.
      10   20   mm

Larvae of the tuber flea beetle make narrow tunnels in tubers. Some of the tunnels turn brown or black in response to fungal invasion.

Infestations are most common in sandy soil where sod or other organic matter has been plowed in before planting. False Japanese beetle larvae are found only in sandy soils and are not associated with sod. They are most damaging to tubers that remain in the soil a long time before harvest, apparently because they move to tubers when other organic matter has dried up or disappeared. Special management practices are seldom needed in potatoes.

Leatherjacket
Tipula dorsimaculata (Tipulidae)

The larvae of certain crane flies, leatherjackets are wormlike, legless maggots with tough, folded skin resembling leather. Mature larvae are gray to brown and up to 1½ inches (4 cm) long. The small, pointed head retracts deeply into the thorax and may not be visible when specimens are handled; the tip of the abdomen bears a group of small fingerlike projections. Leatherjackets feed largely on decaying plant debris but also feed on potato tubers, producing round holes up to 1 inch (2.5 cm) deep. Infestations are rare except in fields following an alfalfa crop plowed under in the spring; the leatherjackets feed at first on the decaying alfalfa, moving to tubers later. Special management practices are not necessary for leatherjackets.

Seedcorn Maggot
Delia platura (Anthomyiidae)

The seedcorn maggot is a widespread pest that occasionally damages potato seed pieces in cool, wet soil. The maggots hatch from eggs laid in moist soil by the adult female, a drab gray fly about the size of a housefly. The legless, whitish maggots, about ¼ inch (6 mm) long when mature, burrow into seed pieces and underground stems; they feed for 2 to 3 weeks before pupating in the soil and emerging as adults. There may be two or three generations a year, but only the first affects potatoes.

Damage is more likely in soil where a lot of plant debris is plowed in before planting. A delay in planting can reduce the chance of injury in such fields by allowing the soil to warm up, reducing the period when seed pieces are vulnerable. Seed pieces that are well suberized are seldom damaged.

Garden Symphylan
Scutigerella immaculata (Scutigerellidae)

Garden symphylans are slender, whitish arthropods related to centipedes. They are found throughout the western United States, but infestations are usually limited to small areas with soils high in organic matter, where they persist from year to year. They are a serious pest in the Pacific Northwest west of the Cascades.

Adults are up to about ⅜ inch (1 cm) long, with 10 to 12 pairs of legs and a pair of antennae. They run rapidly to escape light when exposed in soil. Eggs are laid in soil, and immatures develop to maturity through a series of molts; immatures resemble adults but have only 6 to 10 pairs of legs. Under favorable conditions, adults live for several years. Symphylans move up and down in the soil in response to temperature and moisture. They are most active at 50° to 70°F (10.0° to 21.1°C).

Symphylans feed on roots and other underground structures of many crops and weeds. Damage to roots can stunt

White grubs are usually curled into a C shape with the tip of the abdomen swollen and dark gray. Several larvae of the tenlined June beetle are shown here.

JUNE BEETLE

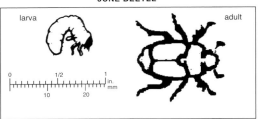

larva

adult

Larvae of the seedcorn maggot are whitish, legless maggots. A larva is shown here (top) with several pupae and a prepupa (left).

potato plants. An injured tuber has small, round entry holes leading to irregular chambers beneath the skin; a ring of dark, corky tissue develops around each entry hole. Symphylans do not produce narrow tunnels such as those caused by flea beetle larvae.

Where treatment is necessary, soil fumigation can protect potatoes for up to 2 years if the treatment is thorough. Continuous flooding for 3 weeks or more in summer also reduces symphylan infestations. More information about garden symphylans can be found in *Pacific Northwest Insect Management Handbook*, listed in the suggested reading.

OTHER INSECTS

Various insects feed occasionally on potato plants and have little or no effect on the crop, but you should be able to recognize them so that you do not confuse them with damaging pests.

Blister Beetles (Meloidae)

Potatoes in fields close to desert or rangeland are occasionally infested with blister beetles. The larvae develop as predators on the eggs of grasshoppers or ground-nesting bees. The adult beetles cluster near the tops of plants, usually in small areas of the field; they feed on leaves and flowers. Treatment is rarely if ever necessary.

Blister beetles are easily recognized by the narrow, neck-like first segment of the thorax. The species found on potato in the western states are the spotted blister beetle (*Epicauta maculata*), the ashgray blister beetle (*E. fabricii*), the punctured blister beetle (*E. puncticollis*), and the Nuttall blister beetle (*Lytta nuttalli*).

Plant Bugs (Miridae, Lygaeidae)

A number of plant bugs may cause minor damage on potatoes. These include lygus bugs (*Lygus elisus*, *L. hesperus*, and *L. lineolaris*, family Miridae) and false chinch bug (*Nysius raphanus*, family Lygaeidae). Feeding by plant bugs may damage young buds and sometimes causes distorted growth or wilted leaves, but generally does not affect yield. Often, the damage is not noticed until the bugs have already left the field. Lygus populations often develop on alfalfa or weeds. When the alfalfa is harvested or the weeds dry out, adult lygus bugs may migrate to crops, including potatoes. False chinch bug populations build up on weed hosts, primarily mustard family weeds, and migrate to other crops when these weeds dry out.

Leafhoppers (Cicadellidae)

In the western states, leafhoppers are of concern in potatoes primarily because they are vectors of *Beet curly top virus* (BCTV) and certain phytoplasmas. The beet leafhopper, *Circulifer tenellus*, occasionally causes problems in potatoes

Garden symphylans are white and run rapidly to escape light when uncovered.

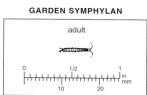

GARDEN SYMPHYLAN

adult

by transmitting BCTV or beet leafhopper transmitted virescence agent (BLTVA). The aster leafhopper (*Macrosteles quadralineatus* = *M. fascifrons*), mountain leafhopper (*Colladonus montanus*), and Flor's leafhopper (*Fieberiella florii*) may transmit phytoplasmas causing aster yellows or witches' broom. It is possible that these leafhoppers may be able to transmit BLTVA, but their role as vectors of this phytoplasma are unknown. Insecticides applied for aphids and other pests normally keep leafhopper numbers low in potato. Applications aimed specifically at leafhoppers may be needed where leafhopper populations are likely to spread BLTVA into potato fields. Check with local authorities for the latest recommendations in your area.

The intermountain potato leafhopper, *Empoasca filamenta*, is the most common of the various leafhoppers found on potatoes in the western states. Adults fly into potato fields after the population has passed through one or more spring generations on weeds. The leafhoppers feed mostly on the undersides of lower leaves, causing small chlorotic spots that give leaves a speckled or stippled appearance. However, this injury does not cause significant damage in potatoes, and this leafhopper does not transmit phytoplasmas. The potato leafhopper, *Empoasca fabae*, can cause serious "hopper burn" damage to potatoes in the Midwest, but this leafhopper does not occur on potatoes in the western states.

Thrips (Thripidae)

Thrips are tiny, slender insects that feed by rasping the surface of leaves. The species found on western potatoes include the western flower thrips, *Frankliniella occidentalis*, and the onion thrips, *Thrips tabaci*. Their feeding causes brown scarring or silvering on the undersides of leaves; severely damaged leaves dry out and drop. The damage often is not noticed until the thrips population has declined or disappeared. Thrips damage is similar to that caused by windburn or blown sand, but you can recognize it by the numerous black flecks scattered over the discolored area; these are the feces of the thrips. Thrips usually build up in weedy areas or nonirrigated pastures, moving to potatoes only when other hosts begin to dry out; damage is usually limited to the outer few rows adjacent to these sources.

The mountain leafhopper can transmit the phytoplasmas that cause aster yellows and witches' broom. A nymph (bottom) and adult are shown here.

Flor's leafhopper also is able to transmit the aster yellows and witches' broom phytoplasmas to potatoes. The nymph (top) and adult are shown here.

The intermountain potato leafhopper and other *Empoasca* species are elongated and green. Feeding by these leafhoppers may cause a yellow stippling of potato foliage but is usually not serious.

The aster leafhopper transmits the phytoplasmas that cause aster yellows and witches' broom in potatoes.

The beet leafhopper can cause problems when it transmits BCTV to potatoes and other crops or when it transmits the phytoplasma that causes beet leafhopper transmitted virescence in potatoes.

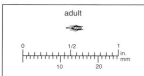

LEAFHOPPER

adult

0   1/2   1
in.
mm
10   20

L. DUNNING

Diseases

Potatoes are susceptible to a large number of diseases. Several occur in all growing areas; some are limited by climate or other factors to one or a few growing regions. Usually, a few diseases are most important in each area. Before you plant, find out which diseases are likely to be problems. Whenever possible, review records of disease incidence in potatoes and other crops that have been grown in the field you intend to plant. Choose cultivars and plan cultural practices that will reduce the impact of key diseases. Always use certified seed tubers and follow good seed-handling practices.

Diseases can be classified as biotic or abiotic depending on the nature of the causal agent. Biotic diseases are caused by pathogens—fungi, bacteria, nematodes, phytoplasmas, viruses, or parasitic plants such as dodder—that invade host plants and disrupt their normal functions. Abiotic diseases (physiological disorders) are caused by physical factors—environmental stresses, toxic substances, or nutrient deficiencies. Physiological disorders are discussed in the next chapter, and nematodes in the following chapter.

To effectively manage biotic diseases, you must know where pathogens originate (primary inoculum sources), how they disperse, how they infect potato plants, and what environmental and cultural conditions favor disease development. Pathogens may be classified by the part of the host plant they affect and their method of dispersal. Root, stem, and tuber pathogens usually are soil-inhabiting or soil-invading organisms that are moved in contaminated soil or infected tubers; they usually infect plant parts in contact with soil. Bacterial and fungal pathogens that attack leaves are dispersed by wind or water and infect aboveground plant parts. Viruses and phytoplasmas are spread mechanically, by insect, water mold, or nematode vectors, or in tubers. Because they are carried in the vascular system, viruses and phytoplasmas infect all parts of the plant.

Tuber diseases include those such as silver scurf, pink rot, and common scab that are limited primarily to tubers. Several diseases that affect other parts of the plant also cause serious tuber diseases. These include late blight, blackleg, ring rot, potato leafroll, necrotic strains of *Potato virus* Y, Rhizoctonia (black scurf), and early blight. In most cases, the use of proper cultural practices reduces losses to tuber diseases. By using certified seed, suberizing seed tubers properly, providing adequate but not excessive soil moisture throughout the growing season, allowing tubers to mature before harvest, handling carefully, and providing correct conditions for curing and storage of tubers, you can help reduce losses to tuber diseases. Planting resistant or tolerant varieties helps reduce losses to common scab and net necrosis caused by *Potato leafroll virus* (PLRV). Certified seed and strict sanitation practices are required to control ring rot. Chemical seed treatments can help control some tuber diseases.

Field Monitoring and Diagnosis

Before the season begins, familiarize yourself with the symptoms of potato diseases common in your area. Study the sections in this manual on specific diseases with their accompanying illustrations; they will help you know what to look for in the field. Because some diseases can spread rapidly, quick diagnosis is very important.

During the season check fields regularly for disease symptoms and signs of stress. Check developing tubers for abnormal growth, damage, or decay. If symptoms of disease or stress appear, note their distribution within the field. Do they appear on only a few scattered plants? Are they concentrated in certain areas? Or are they generally distributed throughout the field? Also keep track of disease levels to assess how quickly diseases are developing. Keep records of these observations, weather, soil conditions, and previous disease outbreaks to help in diagnosis and in future planning. Soil moisture is particularly important because of its influence on tuber development and many potato diseases.

Every plant disease involves a complex interaction between the causal agent, the host plant, and the environment. The type of disease symptoms produced and the rate at which they develop are influenced by the genetic characteristics of both the plant and the pathogen, by the stage of growth when infection or stress occurs, and by environmental conditions, particularly temperature, soil moisture, and humidity within the plant canopy. When comparing field symptoms with illustrations and descriptions in this manual, consider the possible influence of the environment or other factors on symptom expression. Examine as many affected plants as possible and look for different stages of disease to see how symptoms change during disease development. Look at all parts of the plant, including stems, roots, and tubers. Do not rely on a single symptom such as wilting or leaf yellowing to identify a disease; different diseases may produce similar symptoms if the causal agents affect the same plant function. Observation of several different symptoms is usually needed to identify a disease accurately.

Field symptoms are not always sufficient to make an accurate disease identification. If you are not sure of an identification, seek confirmation from qualified professionals. Some pathogens such as viruses or phytoplasmas may require special laboratory techniques for identification; diagnosis of nutrient deficiencies may require plant tissue analysis. However, even when laboratory analysis is required, an accurate set of field notes will help confirm the results.

Prevention and Management

Seed Selection. Planting certified seed tubers reduces or eliminates losses caused by many diseases that are tuber-borne, including late blight, potato viruses, bacterial ring rot, and several other pests such as silver scurf and black scurf. Seed tubers grown under a state's seed certification program are certified to have shown no more than certain low percentages, called tolerances, of pest and disorder symptoms or chemical injury during required inspections. Pests for which tolerances are enforced and the range of those tolerances among the western states are listed in Table 4 in the third chapter of this manual, "Managing Pests in Potatoes." Familiarize yourself with the certification requirements that apply to the seed tubers you buy, and obtain your seed from a source with a reputation for supplying good-quality seed tubers. If possible, visit the areas where the seed tubers are grown to investigate conditions and practices that influence seed quality and look at the general condition of the crop during the growing season and in storage. For the seed lots you are considering, request a North American Certified Seed Potato Health Certificate, which reports the results of winter and summer test readings for that lot and contains its pedigree. Current examples of Seed Potato Health Certificates can be found at the Potato Association of America Web site (www.umaine.edu/paa) and the Potato Information Exchange Web site (oregonstate.edu/potatoes). Follow careful handling and sanitation procedures to reduce contamination of your certified seed tubers.

Cultivar Selection. Potato cultivars vary in their susceptibilities to a number of important diseases and physiological disorders. Choosing cultivars that are the least susceptible to problem diseases is an important component of a good IPM program. Table 6 in the third chapter, "Managing Pests in Potatoes," lists the susceptibilities of commonly grown cultivars to several important diseases and physiological disorders.

Field Selection. Several potato pathogens can remain dormant in the soil for long periods and can cause disease when potatoes are planted again. Some potato pathogens can cause diseases on other crops. Keep records of diseases that occur both in potatoes and in rotation crops as a guide for future management decisions. It is especially important to obtain information about disease histories, particularly in any potato rotations, when buying or leasing ground. Use these records to help make decisions on what cultivars to plant or what special management practices to use. For example, in a field with a history of Verticillium wilt, you may want to avoid planting a highly susceptible cultivar such as Russet Norkotah, you may decide to fumigate, or you may choose to avoid planting potatoes. Histories of powdery mildew or severe early blight indicate a need for special fungicide applications or monitoring during the growing season. You may want to avoid planting potatoes in fields with a history of root knot nematode, potato rot nematode, or deep-pitted scab; if growing seed potatoes, avoid fields that have had problems with Verticillium wilt, corky ringspot, or root knot nematodes.

Avoid fields where poor drainage or other conditions will interfere with uniform irrigation. Anything that makes water management difficult increases the risk of damage by several potato pathogens and disorders.

Cultural Practices and Sanitation. Correct storage, handling, and planting of seed tubers, along with proper management of soil moisture and fertility, minimize losses to most potato diseases. Sanitation is essential to the prevention of seed piece infection during cutting and handling and to the prevention of the spread of pathogens in contaminated soil, water, and field equipment. Strict sanitation requirements must be followed in growing seed potatoes. A list of cultural practices important for controlling diseases and other pests is presented in Table 3 in the third chapter, "Managing Pests in Potatoes." Cultural practices and sanitation procedures are discussed in the sections on individual diseases and in the third chapter.

Pesticides. Fungicides can reduce damage caused by certain foliar pathogens such as late blight, early blight, and powdery mildew. To be effective, they must usually be applied before infection occurs or when the disease just begins to develop. Therefore, it is important to know the history of disease problems in the field, to take into account weather conditions that favor disease, and to monitor for the appearance of disease symptoms or rely on disease forecast models where available. Soil fumigants may control nematodes or Verticillium wilt. Follow the guidelines given in the discussions of individual diseases and consult local authorities for current information on materials recommended to manage disease problems in your area. A general discussion of pesticide use, hazards, and safety procedures can be found in the third chapter, "Managing Pests in Potatoes," and in several references listed in the suggested reading.

ROOT, STEM, AND TUBER DISEASES

Diseases of roots, stems, and tubers are caused by fungi, bacteria, or nematodes that are transmitted in infected seed tubers, plant debris, or infested soil. Nematodes are covered in the seventh chapter of this book. Management of root, stem, and tuber diseases generally requires certified seed to reduce spread of tuberborne pathogens; fungicide treatment and proper suberization of seed pieces to prevent infection; careful irrigation scheduling to avoid excessively wet or dry conditions during critical growth phases; and maintenance of

healthy plant growth. Crop rotation is useful for control of soil-inhabiting pathogens that have limited host ranges and require host plant residues for survival, such as the pathogens that cause bacterial ring rot and silver scurf. Rotation is less effective for pathogens such as *Verticillium dahliae* or *Phytophthora erythroseptica*, which can survive in the soil for a long time in the absence of a host. Resistant or tolerant varieties can help reduce losses caused by some soilborne diseases such as Verticillium wilt and common scab.

Verticillium Wilt
Verticillium dahliae

Verticillium wilt is one of the most serious potato diseases, and it commonly occurs wherever potatoes are grown in the United States. Symptoms usually appear in mid to late season, but they develop sooner and are more severe when plants are stressed. The term "early dying" is frequently used to describe some symptoms of Verticillium wilt, because affected plants appear to be aging or dying prematurely. Verticillium wilt reduces yield; the earlier in the season or more severely plants are affected, the greater the yield reduction. In most growing areas, maintaining plant vigor by using proper cultural practices controls the expression of wilt symptoms and reduces losses. Soil fumigation may be helpful in some situations.

Symptoms and Damage

Symptoms generally first appear on lower leaves. Rarely, symptoms first appear on upper leaves; this is most likely to occur when wilt develops early in the season. Areas between the leaf veins turn yellow and then brown. Affected plants may wilt on warm, sunny days and recover in the evening. Early in disease development, wilting and yellowing may affect leaflets on one side of a petiole or leaves on one side

Leaves of plants affected by Verticillium wilt turn yellow, then turn brown as leaf tissue dies. As plants become severely affected, they tend to stand up higher than other plants in the field, a symptom called flagging.

of a stem. Unilateral wilting is a symptom that usually occurs on one stem of a plant. Wilting can be caused by other diseases, particularly blackleg, Rhizoctonia, or bacterial ring rot; in these cases other symptoms that distinguish them from Verticillium wilt are usually present. Foliar symptoms of Verticillium wilt move up the plant until all the foliage turns brown and dies. As the disease progresses, diseased plants may exhibit flagging—standing more upright than unaffected plants.

After foliar symptoms appear but before they become severe, the water-conducting tissue (xylem) of the lower stem and root turns a reddish to dark brown. This symptom can be seen by splitting the lower stem and root in half to expose the darkened vascular tissue. Vascular discoloration is also caused by blackleg, but the color tends to be grayish brown to black and may not be restricted to xylem. Stems affected by blackleg (bacterial stem rot) are softened and usually turn black, symptoms not associated with Verticillium wilt. Some xylem discoloration may be seen in plants wilting from bacterial ring rot.

Vascular discoloration caused by Verticillium infection may extend into the stem ends of tubers on some cultivars. However, Verticillium may infect tubers without causing symptoms. PLRV or physiological stress may cause similar tuber symptoms.

The fungus interferes with transport of water and nutrients in the xylem, so symptoms are most severe on coarse-textured soils and during periods of hot weather, when plants are more easily stressed for water. Under cool conditions and when soil moisture and fertility are kept at adequate levels, symptoms may be less evident. Certain species of root lesion nematode, *Pratylenchus penetrans* and *P. thornei*, can increase the severity of Verticillium wilt. Verticillium wilt is often more severe in

areas such as the Columbia Basin where the season is long, temperatures are extreme, and soils are coarse.

Seasonal Development

Verticillium dahliae is a soil-inhabiting fungus that has many different strains causing wilt diseases in tomatoes, peppers, eggplant, melons, cotton, stone fruits, and many other crops and wild plants (Table 13). In other parts of the United States and elsewhere in the world, *V. albo-atrum* causes wilt of potatoes and other crops. In the absence of a susceptible host, *V. dahliae* can survive in the soil for 8 years or more as tiny black structures called microsclerotia. Microsclerotia are formed in dead or dying stems of infected plants and are the source of inoculum for future infections.

In the presence of a susceptible host plant, the microsclerotia germinate. The fungus invades the roots through the region of cell elongation located behind the root tip. The fungus establishes itself in the xylem elements, where it spreads upward into stems, petioles, and leaflets.

The water-conducting tissue (xylem) of stems infected by *Verticillium dahliae* turns reddish or dark brown. To see this symptom, slice or peel the lower stem and upper root in half to expose the vascular tissue.

Table. 13. Susceptibility of Several Crop Plants to *Verticillium dahliae* from Potato.

Crop	Susceptibility	Comments
alfalfa	highly susceptible	
artichoke	highly susceptible	
bell pepper	susceptible	
broccoli	highly resistant	Rotation with incorporation of crop residue reduces Verticillium wilt inoculum in the soil.
cabbage	susceptible	
cauliflower	susceptible	
chili pepper	highly susceptible	
corn	highly resistant	Corn rotations reduce Verticillium wilt inoculum in the soil.
cotton	highly susceptible	
eggplant	highly susceptible	
grains	highly resistant	
lettuce	highly susceptible	
mint	resistant	Isolate that attacks potato can cause mild symptoms on mint, and mint can maintain soil populations of this isolate.
mustards	highly resistant	Incorporation of cover crop residue reduces Verticillium wilt inoculum in the soil.
strawberry	highly susceptible	
sudangrass	highly resistant	Incorporation of cover crop residue reduces Verticillium wilt inoculum in the soil.
tomato	moderately susceptible	
watermelon	highly susceptible	

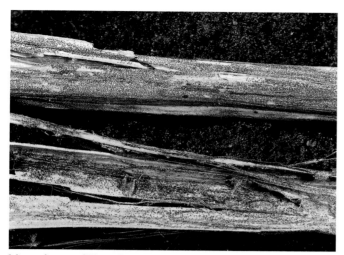

Microsclerotia of *Verticillium dahliae* are tiny black structures formed in dead and dying stems of infected plants. They can survive for long periods in the soil and are the source of inoculum for future infections.

As it spreads through the vascular system, the fungus impedes water movement and causes browning of the tissues. Growth of *Verticillium* in the plant is favored by cooler temperatures, about 55° to 75°F (13° to 24°C). Foliar symptoms may appear once *Verticillium* is established in a plant, although the fungus is frequently present in healthy-looking potato plants without causing symptoms. More-apparent symptoms are evident when infected plants are stressed by insufficient irrigation, deficient nitrogen, or other diseases. For that reason symptoms are earlier and/or more prominent on sandy ridges.

The fungus is present in most soils where multiple crops of potatoes have been grown. Inoculum levels increase each time a crop of infected potato vines is plowed under. Infected seed potatoes are thought to be the primary source of *Verticillium* in new ground. Dispersal of inoculum within fields is primarily by cultivation. Some movement between fields may occur in windblown soils or on contaminated equipment.

Management Guidelines

Resistance to Verticillium wilt varies among potato cultivars. Bannock Russet and Chipeta are resistant; Atlantic, Ranger Russet, Umatilla Russet, and White Rose are moderately resistant; Russet Norkotah, Shepody, and Yukon Gold are susceptible. Most cultivars express a tolerance to this disease when proper soil fertility and moisture are maintained. When *Verticillium* inoculum in the soil is high and conditions are favorable for disease expression, even the most resistant commercial cultivars may develop symptoms.

Maintaining plant vigor with good fertilization and irrigation practices is the best cultural management presently available for Verticillium wilt. Proper soil fertility suppresses development of symptoms and reduces losses, especially in indeterminate cultivars; adequate nitrogen, phosphorus, and potassium are particularly important. Irrigating with sprinklers rather than furrows makes it easier to apply precise amounts of nitrogen and water. Overirrigation earlier in the season prior to tuber set has been shown in the Columbia Basin to increase levels of Verticillium wilt. In some cases, all plants in a field may be infected with *Verticillium* without showing symptoms until they are stressed.

Because *Verticillium* microsclerotia survive in the soil for 8 to 14 years, rotations of at least 5 years are needed to reduce the amount of *Verticillium* inoculum in the soil. However, rotating into field corn or legume crops other than red clover for 2 or 3 years does prevent the level of *Verticillium* in the soil from increasing. Recommended rotation crops include corn, sugar beets, onions, and peas; late plantings of Sudangrass following grain or peas are also useful. Working-in the residue of some cole crops and mustard cover crops has been shown to reduce *Verticillium* inoculum levels substantially. Although *Verticillium* may survive on the roots of grain crops, grain rotations are better than no rotations at all. Table 13 lists the susceptibility of a number of crops and cover crops to *Verticillium dahliae*. Ask your local extension agent, farm advisor, or other experts for recommended rotation crops in your area. Control weeds and volunteer potatoes in rotation crops, because they may be hosts for *Verticillium*.

Soil fumigation with metam sodium is cost effective for coarse-textured soils where wilt is serious. However, *Verticillium* can build back to damaging levels after one season of potatoes following soil fumigation. Fumigation is more effective in coarse-textured soils; it is usually not economical for medium or fine-textured soils. To determine if soil fumigation is necessary for a particular field, test soil to determine the level of *Verticillium dahliae*. The presence of levels above 10 cfu per gram of soil in the Columbia Basin may cause substantial yield reductions. The potential for loss is also related to cultivar, soil fertility, and irrigation practices. Fumigate a test strip to see how it affects yields. The root lesion nematodes *Pratylenchus penetrans* and *P. thornei* may increase the severity of Verticillium wilt. In some areas, nematicide treatments will reduce wilt levels.

Sanitation of field equipment to avoid introduction of contaminated soil and burning or effective removal of vines before harvest are especially important in fields that are planted to potatoes for the first time. Seed tubers destined for new fields should be tested for the presence of *Verticillium* and not planted if infected. Complete burning of vines is a very effective way to reduce the introduction of new *Verticillium* inoculum to the soil, because infected stems are the main source of microsclerotia. However, burning is not used routinely, primarily because of cost. If you are considering burning vines, check with local regulations to be sure the practice is allowed.

Blackleg, Aerial Stem Rot, Tuber Soft Rot

Pectobacterium carotovorum (*Erwinia carotovora*)

Blackleg, aerial stem rot, and tuber soft rot occur wherever potatoes are grown. The symptoms expressed and severity of the disease depend on seed-handling techniques, soil moisture and temperature at planting, environmental conditions, cultivar, amount of infection in the seed lot used, and external sources of the bacteria such as irrigation water. Blackleg can be a major disease problem if more-susceptible varieties such as Norland and Shepody are grown. The disease is best controlled by warming seed tubers before cutting, making sure seed temperatures are not lower than soil temperature at planting, and planting healthy seed pieces when soil is at least 50°F (10.0°C) and moist but not wet. Careful selection of seed stocks and good sanitation during seed cutting help reduce the incidence of blackleg.

Symptoms and Damage

Blackleg symptoms are caused by bacterial inoculum present in the seed piece and may appear at any time during the season. If warm, wet soil conditions occur soon after planting, infected seed pieces will decay entirely before emergence, resulting in reduced stands. When young plants are affected, stunting and erratic growth may occur. On older plants, yellowing between leaf veins and browning and upward curling of leaf margins are the first symptoms.

Slicing or peeling stems in half should show a grayish brown to black discoloration that is not confined to vascular tissue but sometimes spreads into the pith. As this discoloration progresses, stems may turn an inky black and become soft and mushy just above and below the soil line. This inky black symptom, known as "blackleg," usually occurs during the course of the disease, but sometimes stems will become soft and mushy without the black discoloration.

Under some conditions plants may appear stiff and erect. Occasionally aerial tubers form. Leaf curling, yellowing, and aerial tubers may also be caused by Rhizoctonia, phytoplasmas, or waterlogged soils. Infected plants may wilt during hot weather. Entire stems may wilt without discoloration of leaves or wilting may occur with yellowing and browning of leaves, similar to Verticillium wilt. Wilting of plants later in the season due to bacteria has been termed "bacterial early dying;" stems of affected plants have an internal brown discoloration but no blackleg symptoms. In some cultivars, wilting and leaf symptoms may appear similar to symptoms caused by bacterial ring rot, but tubers of plants affected by blackleg do not show the vascular decay seen with ring rot.

Aerial stem rot is caused by aboveground sources of inoculum rather than bacteria that move up into stems from infected seed pieces. Symptoms appear as water-soaked lesions on stems that ultimately turn light brown or, occasionally, black.

Tuber soft rot is caused by the same bacteria that cause stem symptoms, but it is not always associated with plants that show blackleg symptoms. Tuber symptoms range from a

Stems of plants with blackleg usually turn black just above and below the soil line. Plants may wilt and leaves may curl up and turn brown at their margins.

Plants infected by *Pectobacterium* (*Erwinia*) may wilt without developing blackleg symptoms.

Gray-brown discoloration of the vascular tissue develops in stems invaded by *Pectobacterium* (*Erwinia*) that has spread from decayed seed pieces. The discoloration does not remain confined to the vascular tissue, as in the case of Verticillium wilt, but spreads into the pith.

GARY D. KLEINSCHMIDT

Bacterial soft rot lesions caused by *Pectobacterium* (*Erwinia*) are separated from undecayed tissue by dark brown or black margins. The lesions may later turn hard, black, and dry.

slight discoloration of the vascular tissue to complete soft rot of the tuber. Soft rot lesions develop most often at wounds, in lenticels, or where stolons attach. Lesions at wounds usually begin as light-colored, odorless, soft decay, sharply separated from undecayed tissue by dark brown or black margins. Lesions around infected lenticels appear sunken, water-soaked, and circular in shape. Both types of lesions can turn hard, black, and dry in storage. Other bacteria often invade the lesions, causing a soft, foul-smelling decay. The amount of tuber soft rot is usually greater when certain dry rot fungi are present, particularly *Fusarium*. When tubers are infected by the late blight, pink rot, or Pythium leak pathogens in the field, they are also invaded by soft rot bacteria. Ultimately soft rot bacteria are responsible for wide-scale losses that occur in potato storages.

Seasonal Development

Two variants or subspecies of the bacterium *Pectobacterium carotovorum* (= *Erwinia carotovora*), *P. c.* subsp. *atrosepticum* and *P. c.* subsp. *carotovorum*, cause blackleg, aerial stem rot, and soft rot of tubers. The former is more active under cooler temperatures, below 70°F (21°C), and in some situations causes more of the inky black symptom associated with blackleg. The latter is more active under warmer temperatures, above 70°F, and usually causes more stem rot without the inky black symptom. *P. c.* subsp. *carotovorum* and *P. chrysanthemi* (= *Erwinia chrysanthemi*) cause wilting of plants late in the season (bacterial early dying). These bacteria survive in infected tubers, in surface water sources, on the roots of other crops including grains and sugar beets, and on the roots of a number of weeds common in potato fields, including nightshades, lambsquarters, pigweeds, Russian thistle, kochia, purslane, and mallow.

Pectobacterium bacteria reside in lenticels without causing decay, but they may cause blackleg or early dying symptoms if tubers are used for seed. The bacteria may be spread from diseased to healthy seed pieces during cutting. Infection of seed pieces is most likely if wound healing is delayed, usually because tubers are not warmed before cutting or because warm seed pieces are planted in cool, dry or excessively wet soil. Warm soil favors wound healing, but if the soil is wet, infected seed pieces will decay before any sprouts emerge. If the soil is cool after planting, infected seed pieces decay more slowly and the bacteria slowly move up into the water-conducting tissue of the stems. The bacteria multiply and degrade the vascular tissues, causing foliar symptoms and wilt. Later, bacteria spread into stem pith tissue, causing the darkening and softening that are the typical blackleg symptoms. In some cases, particularly in the hot growing areas of the low-elevation desert valleys, seed pieces decay later in the season and bacteria spread up into stems, destroying the vascular tissue and causing plants to wilt and die without typical blackleg symptoms.

Bacteria in surface irrigation water may be an important inoculum source for infection of leaf petioles and stems through wounds or lenticels. Decaying seed pieces are a source of inoculum for contamination of daughter tubers. Spread of bacteria from plant to plant and tuber to tuber is favored by wet conditions. Splashing rain or irrigation water spreads bacteria to aboveground plant surfaces, where infection results in aerial stem rot.

Wet soil conditions favor tuber infections; immature tubers are more susceptible to bacterial infection than are tubers with fully developed skin. Tubers infected with late blight, pink rot, or Pythium leak are more susceptible because lesions of these diseases can be infected by bacteria. Proper curing and a low storage temperature prevent or restrict development of soft rot symptoms. Moisture leaking from soft-rotted tubers spreads infection in storage. Free moisture on tuber surfaces and poor storage ventilation restrict the oxygen supply to tubers and favor spread of the disease because the bacteria can multiply under low oxygen and the resistance of tubers is compromised. If tubers with soft rot are placed in storage, adequate airflow with reduced humidity should be used to dry infected potatoes and keep the soft rot bacteria from spreading.

Management Guidelines

Using seed stocks with low levels of the pathogen is a fundamental part of managing seed piece decay and blackleg. Obtain seed tubers from a reputable source, such as seed originating from stem cutting or micropropagation programs and grown under a limited-generation program. However, poor handling or conditions conducive to disease after planting can result in high amounts of blackleg or seed piece decay, even when the best-quality seed is used. Shipping point inspections by qualified personnel can be useful in assessing levels of bacteria in given seed lots.

To reduce the spread of bacteria during seed cutting, hold seed tubers at 50° to 55°F (10.0° to 12.8°C) with relative humidity at 95% for 10 to 14 days before cutting, if excessive sprouting is not a problem. Follow good sanitation practices during seed cutting and handling and keep cutting blades properly sharpened. Wash equipment clean and apply the recommended rate of a registered disinfectant (see Table 7 in the third chapter, "Managing Pests in Potatoes") at least twice a day and always between seed lots. Treat seed pieces with a recommended fungicide during or after cutting.

Plant in well-drained, moist soil with a temperature of 50°F (10.0°C) or higher to encourage rapid wound healing. Colder and drier conditions slow wound healing, increasing the likelihood of infection. Never plant seed pieces with a pulp temperature lower than the soil, because moisture from the soil will condense on the surface of the seed. Avoid excessive soil moisture, which favors rapid seed piece decay. When planting cannot be done in warm soils, treatment of seed tubers with approved fungicides may reduce seed piece decay, because some seed treatment materials control *Fusarium*, which interacts with *Pectobacterium* to cause more serious seed piece decay.

When growing seed potatoes, avoid using surface water for irrigation. Remove plants with blackleg or aerial stem rot symptoms as soon as they appear. If you rogue blackleg plants, be sure to remove all tubers; place them in plastic bags as soon as you pull them from the soil to avoid spreading bacteria.

Because immature tubers are infected more easily due to increased mechanical injury, allow tubers to mature before harvest. Avoid excessive soil moisture at harvest. Properly suberize tubers after harvest by holding them at 50° to 55°F (10.0° to 12.8°C) and 95% relative humidity for 10 to 14 days with good ventilation. Reduce pulp temperatures of stored potatoes as soon as possible, while allowing for healing, and ensure good ventilation. Recommended storage temperatures can be found under "Storage" in the third chapter, "Managing Pests in Potatoes."

Studies have shown that increasing calcium nutrition can reduce tuber soft rot. Research is underway to see if increased rates of calcium fertilizer in seed potato fields can reduce the incidence of seed piece decay and bacterial early dying when tubers from these fields are used for seed.

Rhizoctonia Stem and Stolon Canker and Black Scurf
Rhizoctonia solani

Though widespread, Rhizoctonia stem and stolon canker causes only small yield reductions in the West unless cool, wet weather occurs during sprout growth. Otherwise, the principal damage caused by *Rhizoctonia solani* is loss of tuber quality in fresh market cultivars from black scurf. Planting seed tubers free of *Rhizoctonia* sclerotia, using seed treatments that are effective against *Rhizoctonia*, and using cultural practices that favor rapid sprout emergence are the best control measures.

Symptoms and Damage

Rhizoctonia causes the most severe losses under cold, wet soil conditions during early sprout growth when it can kill sprouts before they emerge. This type of damage is not common in the West, but results in reduced stands and large yield losses.

In the western states, damage is normally first seen after plants are established. Reddish brown lesions on stolons and underground parts of the stems are characteristic. Some infected stolons may be killed as they form. This stolon pruning results in fewer but larger tubers on each plant.

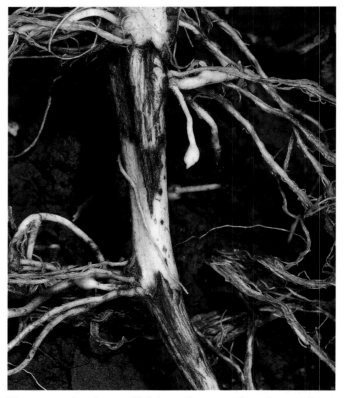

Rhizoctonia solani forms reddish brown lesions on the underground parts of stems.

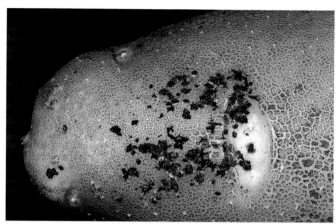

Rhizoctonia solani forms dark brown or black sclerotia, called black scurf, on tuber surfaces.

Sometimes *Rhizoctonia* lesions penetrate stem tissue far enough to interfere with nutrient flow in the conducting tissues. When this happens, the foliage may be wilted, stunted, rosetted, or curled upward, and aerial tubers may develop. Leaves of whiteskinned or russet varieties may turn yellow; leaves of redskinned varieties may turn reddish or purple. Foliar symptoms may appear similar to those caused by PLRV. Aerial tubers may form at the base of stems, but aerial tubers may also be caused by blackleg, phytoplasmas, or waterlogging.

Sometimes a sparse, white growth of *Rhizoctonia* mycelium develops on stems just above the soil line. This growth does not damage the potato plant. Another symptom caused by *Rhizoctonia* is much more common and can cause significant losses in fresh market potatoes. Black bodies, which are survival structures of the fungus called sclerotia, form on the surface of tubers. This symptom, called "black scurf," occurs when soil temperatures are mild and soil is moist, once tubers have formed.

Seasonal Development

Many strains of *Rhizoctonia solani* exist and infect a wide variety of plants including grains, legumes, and vegetable crops. The strain that attacks potatoes does not cause disease on other plants. It survives on plant debris in the soil or as black scurf on the surface of tubers. For unknown reasons, the potato strain of *R. solani* does not survive well in the soil of many areas.

In the presence of growing stems or stolons, the fungus is stimulated to form infection structures. If the soil is wet and cool and the shoot tips are growing slowly, the fungus will completely invade and kill them. Maximum disease development occurs at soil temperatures below 54°F (12.2°C); infections and damage by *Rhizoctonia* decrease at higher temperatures. The fungus will form black scurf on tuber surfaces anytime the soil is moist and cool.

Management Guidelines

Good cultural practices that favor rapid emergence are the best control for Rhizoctonia. Avoid planting in wet or cold soil. Warm the seed tubers before cutting, plant seed pieces when soil temperature is above 45°F (7.2°C), plant shallowly or drag off hills before stem sprouting, and form up hills after sprouts have emerged. Do not irrigate cold soil until sprouts have emerged.

Use crop rotations to keep Rhizoctonia from increasing to more damaging levels. Allow crop residues to decompose before planting; damage may be more severe following sugar beets, alfalfa, or clover if the crop residue is not decomposed before potatoes are planted.

Control of seedborne inoculum is important in potato fields where the potato strain of *Rhizoctonia* does not survive in the soil. Avoid using seed tubers that have *Rhizoctonia* sclerotia on their surface. Seed treatment fungicides are available that effectively control *Rhizoctonia.*

Bacterial Ring Rot
Clavibacter michiganensis subsp. *sepedonicus*

Ring rot is probably the most feared disease of potatoes; there is a zero tolerance for it in seed certification programs, meaning that one plant or tuber diagnosed as having ring rot, either by the presence of symptoms or with serological methods, causes an entire seed lot to be rejected for certification. Because of strict certification requirements, ring rot appears only occasionally in commercial fields. When it does occur it usually causes large yield and storage decay losses. Ring rot is controlled by using certified seed and following strict sanitation practices. Avoid using borrowed equipment whenever possible.

KENNETH W. KNUTSON

In Russet Burbank and a few other cultivars, the terminal leaflets of young stems may appear bunchy or rosetted if infected by bacterial ring rot.

Ring rot causes interveinal tissue of leaflets to turn pale, then brown. Affected stems wilt.

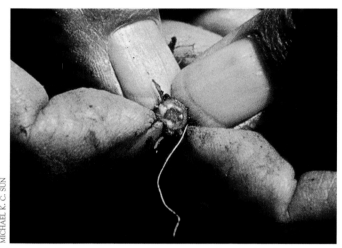

MICHAEL K. C. SUN

Bacterial slime may be squeezed from stems infected with ring rot bacteria.

The vascular ring of a tuber infected with ring rot usually turns brown and corky as it decays. Soft, decayed tissue often can be squeezed out of cross-sectioned tubers. Secondary invasion by other bacteria may cause complete disintegration of tuber tissue.

Symptoms and Damage

In a few cultivars, including Russet Burbank, bunchy or rosetted terminal leaves may appear on young plants. These symptoms are sometimes called "early dwarf" or "green dwarf." Plants with these symptoms usually develop more-typical foliar symptoms of ring rot later.

Foliar symptoms generally appear at midseason or later, usually on middle and lower leaves first. Pale green areas appear on leaf margins or as interveinal mottling and later turn brown, giving leaves a burned appearance. Symptoms move upward until the entire stem wilts and dies. Usually only one or two stems in a hill are affected. Stems sometimes wilt without any discoloration of leaves. In cooler areas, symptoms are less pronounced or may not be expressed; plants may wilt without dying. Symptoms are also masked when soil nitrogen levels are high.

Vascular tissue in stems may turn brown. If infected stems are cut near the seed piece, a milky exudate can usually be squeezed from the cut end.

The characteristic symptom of ring rot is a discoloration and subsequent decay of the vascular ring in tubers. This decay usually appears after foliar symptoms, but it does not always develop in the tubers of infected plants. Tuber decay usually starts as a softening and creamy yellow discoloration of the vascular ring that can be squeezed out after tubers are cut in cross-section. The vascular ring usually turns brown as it decays; browning of the vascular ring sometimes occurs without the soft decay.

Tuber surfaces may become slightly pink or brown, and cracks (sometimes called "chicken tracks") that penetrate to the rotted vascular ring may appear on the surface. Secondary invasion by other bacteria may cause a complete breakdown of the middle of the tuber with little or no discoloration of the vascular ring. Secondary breakdown can occur in the field, in transit, or in storage.

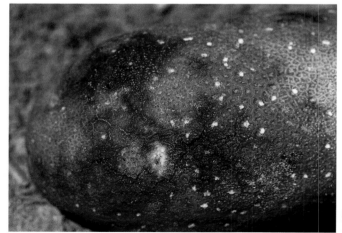

LARRY L. STRAND

The surface of a tuber decaying with ring rot turns pink or brown and cracks develop.

Under cool conditions plants may develop few symptoms while producing infected tubers. Infected tubers can be symptomless at harvest and take up to 2 or 3 months to develop ring rot in storage; sometimes symptoms do not develop in infected tubers.

Seasonal Development

The bacterium that causes ring rot, *Clavibacter michiganensis* subsp. *sepedonicus*, overwinters in infected tubers. It does not live freely in the soil, but it can survive for several years as a dried slime on harvesting and grading machinery, sacks, and surfaces, including earthen floors, of storage and transport facilities. *Clavibacter* is highly contagious; the bacterium infects tubers through wounds that reach into the vascular ring. The principal means of spread is seed cutting; a knife that has cut one infected tuber can spread the bacteria to at least 20 more cut surfaces. Pick planters will also spread ring rot bacteria during planting.

The ring rot bacterium is a vascular parasite. As stems grow from an infected seed piece, bacteria move up in the water-conducting tissues (xylem), multiply, and produce toxins that cause the foliar symptoms. Rosetting and other early dwarf symptoms occur when bacteria proliferate in very young stems of certain varieties.

About one-half to three-fourths of the daughter tubers of an infected plant will be infected with ring rot bacteria. Not all daughter tubers develop ring rot symptoms. Ring rot develops in tubers most rapidly at 64° to 72°F (18° to 22°C), and only slightly at 37°F (2.8°C). Ring rot normally does not spread from tuber to tuber in storage, but pockets of soft rot affecting several tubers may develop around tubers that have disintegrated from ring rot.

Management Guidelines

Use only certified seed tubers. Do not use tubers for seed if they came from an infected lot or a lot that was suspected of having any ring rot. Use only seed tubers free from bacterial ring rot for seed production.

Clavibacter can overwinter in tubers left in the ground, so do not plant potatoes on ground that produced infected potatoes the previous year. Leaving a field fallow or rotating it for at least 1 year should eliminate any ring rot bacteria present, as long as volunteers are controlled. Destroy potato volunteers and cull piles near seed fields.

Clean and disinfect all equipment, bins, containers, sacks—everything that touches tubers. Sanitation is most important during cutting and planting. Keep all seed lots separated from each other and from other potatoes. Clean and disinfect equipment between the handling of each lot or cultivar. All surfaces must be cleaned of debris with detergent and water or steam and treated with a disinfectant registered for ring rot control. Even equipment that has been exposed to the weather or has not been used for potatoes for several years must be

disinfected. Be sure to use correct disinfectant concentrations and treatment times (see Table 7 in the third chapter).

Disinfect storage facilities. Remove all sacks and debris; if storages have dirt floors, remove and replace 2 or 3 inches of soil before disinfecting. Burn contaminated sacks.

If a commercial field shows ring rot, leave tubers in the ground as long as possible to allow most of the infected tubers to disintegrate completely. Harvest the remaining tubers and market them immediately. If you must store them, store them at temperatures as low as possible depending on their intended use, and sell them as soon as possible.

Common Scab
Streptomyces spp.

The causal organism of common scab, a widespread soilborne disease, can infect fleshy roots and underground stems of some other crops and weeds, but it causes economic damage only on potatoes, carrots, and red beets. The surface blemishes it causes do not reduce yields but can reduce the value of a potato crop considerably, particularly fresh market crops. Severity of common scab can be reduced by using resistant cultivars, maintaining adequate soil moisture during early tuber development, using certain soil treatments, and avoiding heavily infested soils.

Symptoms and Damage

The surfaces of tubers infected with scab have brown, roughened, irregularly shaped areas that may be raised and warty, level with the surface, or sunken. Scab lesions may affect just a small part of the tuber surface or almost completely cover it. Three types of symptoms occur: russet scab produces lesions that are superficial corky areas; raised scab causes erupting or cushionlike lesions; deep-pitted scab produces dark brown to almost black lesions that are sunken up to ¼ inch (6 mm) deep. The type and severity of symptoms are influenced by the strain or species of scab organism. Tubers with raised or

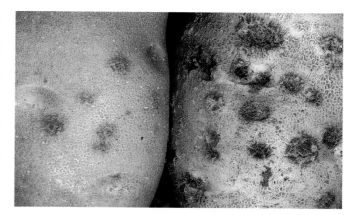

Scab lesions may appear as eruptions, pits, or superficial layers of corky tissue on the tuber surface.

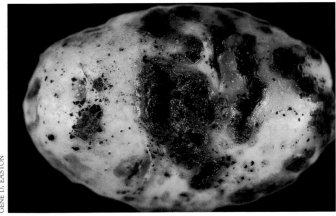

GENE D. EASTON

Lesions of deep-pitted scab are more severe and penetrate more deeply than other types of scab lesions.

russet scab can be used for processing; tubers with deep-pitted scab cannot. Scab tends to be more severe in newly cropped fields and fields high in undecomposed organic matter.

Seasonal Development

Scab is caused by species of the soil microorganism *Streptomyces*. *Streptomyces scabiei* causes russet and raised scab; different species cause deep-pitted scab. *Streptomyces* survives in soil in the absence of host plants and can attack the fleshy roots of weeds and root crops including potatoes, red beets, sugar beets, and carrots. The microbe invades potato tubers through lenticels during the first 5 weeks of tuber development. If tubers dry out during this period, bacteria antagonistic to *Streptomyces* that are normally present in the lenticels disappear, allowing the scab organism to infect more easily. A soil pH of 7 is optimum for scab development; a pH lower than 5.5 usually inhibits it. There is a type of scab called acid scab that develops at a pH below 5, but it occurs on potatoes in the eastern United States and is not known to occur in the western states.

As tubers grow, lesions gradually develop into characteristic scabs. The type of lesion formed depends on the tolerance of the potato cultivar, the species of *Streptomyces* involved, and the soil environment. Lesions do not continue to develop in storage.

Management Guidelines

Potato cultivars differ in their resistance to scab. Russet cultivars tend to be more resistant than smooth-skinned cultivars. Russet Burbank is generally considered tolerant. The relative tolerance of cultivars to scab varies with the area in which they are grown, due to differences in the strains of the disease organisms present and, possibly, environmental effects. If you are planting in fields where scab is a problem, ask local extension agents, farm advisors, or other experts for cultivar recommendations.

Proper soil moisture during tuber development reduces the severity of scab and usually controls the disease adequately. Maintain available soil moisture at 80% of field capacity or more during tuber initiation and early tuber growth, until tubers are golf ball sized. Avoid planting potatoes in fields with severe scab problems. Preplant treatment of such fields with materials designed to lower soil pH or control *Streptomyces*, combined with proper water management, may reduce the amount of scab. Water management and soil treatments may not be effective controls for deep-pitted scab in some areas. If this is the case in your area, you may not be able to grow potatoes profitably in fields where this type of scab occurs.

Silver Scurf
Helminthosporium solani

Silver scurf can be a serious problem in most potato-growing areas. The disease affects only tubers, increasing weight loss during storage and reducing market quality by degrading cosmetic appearance. Silver scurf is managed by using certified seed tubers free of the disease, seed treatment fungicides, crop rotation, sanitation of storage facilities, and maintaining storage conditions least favorable to disease spread.

Symptoms and Damage

Symptoms usually appear first at the stem end as small light brown or grayish leathery spots and may enlarge to cover most of the tuber surface. Lesions have a shiny appearance that is more pronounced when tubers are wet. Margins of young lesions are frequently dark brown and sooty from spore production. Microscopic examination reveals tiny "Christmas tree–like" spore structures. Black dot tuber lesions have a similar appearance but do not have dark, sooty margins and usually contain numerous tiny black dots (sclerotia) that have small microscopic spines. The two diseases

PHILIP B. HAMM

Silver scurf appears as light brown or grayish leathery or russeted areas on the surface or red- or white-skinned tubers. Lesions frequently become dark and sooty from spore production by the pathogen.

When examined under a microscope, the spore-bearing structures of the silver scurf pathogen have a "Christmas tree" appearance.

PHILIP B. HAMM

can be present on the same tuber. Silver scurf is most noticeable on red-skinned and other smooth-skinned cultivars, but can cause substantial damage on russet cultivars. Affected tubers are more susceptible to decay and shrinkage from water loss during storage.

Seasonal Development

The fungus that causes silver scurf, *Helminthosporium solani*, survives on infected seed tubers and on plant debris in infested soil. Spores are produced in lesions that develop on infected seed pieces and spread the disease to daughter tubers. New tubers also can be infected by spores present in infested soil.

The silver scurf pathogen infects tubers through lenticels or directly through the skin and remains confined to the periderm and outer layers of cortex cells. The incidence and severity of the disease increase over time if tubers are left in infested soil. Once tubers are placed into storage, silver scurf lesions can continue to develop, and secondary spread of the disease to other tubes in storage can continue when temperatures are above 37°F (2.8°C) and humidity is above 90%. Disease incidence increases the longer tubers are held in storage. If early-generation seed tubers are held in common storage with later-generation seed tubers, silver scurf may spread to the early generation from the later generation, which is more likely to have some disease present. Likewise, if multiple lots of potatoes are placed into common storage, a lot infected with silver scurf will produce spores that move through the air system, infecting other lots.

Management Guidelines

To minimize silver scurf, plant certified seed tubers free of the disease and use seed treatments known to be effective against the silver scurf pathogen. Harvest as soon as tubers have matured after vinekill. Use humidified airflow to remove free moisture from wet tubers as quickly as possible, and keep storage temperatures as low as possible. Do not plant potatoes in fields that had silver scurf the previous season. Thoroughly clean and disinfect storages before storing a new crop. If possible, avoid storing early-generation seed tubers in the same storage as later-generation seed tubers. Do not repeatedly enter storages over a long period of time and do not begin to remove tubers without completely emptying the storage. Activity in storages encourages the release of spores into the air system and increases levels of silver scurf, particularly if new infections are given enough time to develop. The use of chemicals in the air system has not been proven to reduce losses to silver scurf.

Black Dot
Colletotrichum coccodes

Black dot is a widespread potato disease in western growing areas that may cause substantial losses in some areas. The disease increases losses from early dying caused by *Verticillium* or *Pectobacterium* (*Erwinia*). Black dot also occurs on other plants in the potato family, including tomatoes, eggplants, and peppers. Black dot is managed by using certified seed tubers and crop rotation.

Symptoms and Damage

The disease first appears in mid to late season as a yellowing and wilting of plants. These symptoms are easily confused with Verticillium wilt or bacterial early dying. Wilting caused by black dot develops rapidly, in contrast to Verticillium wilt.

The black dot pathogen can cause severe decay of the cortical tissue of roots. Affected roots may appear stringy when

The black dot fungus forms a thin layer of white mycelium that produces tiny, black, dotlike sclerotia and spore-forming acervuli.

pulled from the soil. In some cases lesions on belowground stems and stolons may be confused with Rhizoctonia stem and stolon canker; however, black dot lesions are darker.

Black dot is distinguished by small, black, dotlike fungal structures (sclerotia) that form on the surface of infected stems, stolons, and tubers. Spore-forming structures (acervuli) with conspicuous hairs (setae) also are formed. The surface layer of infected tubers often turns gray, closely resembling silver scurf. Presence of sclerotia distinguish black dot from silver scurf. Another common symptom of black dot is the adherence of stolons to the stem ends of tubers.

The black dot pathogen often invades plants that are weakened by other diseases, and it may accelerate early death of vines infected with *Verticillium*, *Pectobacterium* (*Erwinia*), and possibly *Phytophthora*. Black dot occurs most frequently on plants grown in coarse-textured soils under conditions of low or excessively high nitrogen, high temperature, or

The spore-forming structures (acervuli) produced by the black dot fungus are black with prominent black hairlike structures (setae).

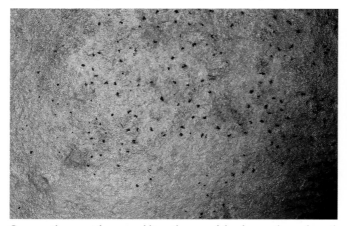

Gray or silvery patches resembling silver scurf develop on the surface of tubers affected by black dot. Presence of the small black sclerotia produced by *Colletotrichum coccodes* distinguish black dot.

poor soil drainage. Tubers of russet cultivars are less severely affected.

Seasonal Development

The fungus that causes black dot, *Colletotrichum coccodes*, survives as sclerotia that form on tubers, stolons, roots, and stems at the end of the season. The pathogen is spread on contaminated seed tubers, and inoculum levels increase in fields continuously cropped to potatoes. Sclerotia can survive for long periods on plant debris in the soil. Spores produced in acervuli on aboveground plant parts can be spread by wind to other plants and cause infections if wounds are present. Black dot infections are increased by the abrasion caused by windblown sand.

Management Guidelines

To manage black dot, plant certified seed tubers, maintain adequate levels of nutrients, and avoid overirrigation. Where fields become infested, rotate to nonhost crops such as grains for at least 3 years and control potato volunteers and potato family weeds in the rotation crops. No potato cultivars are resistant, but early-season cultivars may escape some damage in infested fields, and the tubers of russet cultivars are less severely affected than tubers of thinner-skinned cultivars.

Fusarium Dry Rot
Fusarium spp.

Losses from dry rot occur wherever potatoes are stored but can usually be kept to a minimum with proper handling and storage techniques.

The first signs of dry rot are small brown areas that appear around wounds about 1 month after tubers are put into storage. The skin over the dry-rotted areas sinks and becomes wrinkled; internally, lesions are light to dark brown, dry and

When the black dot fungus destroys the cortical tissue of stolons, a portion of affected stolons remains attached to the tuber, a symptom called "sticky stolon."

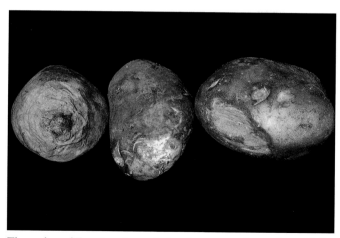

The surface of Fusarium dry rot lesions becomes sunken and wrinkled. White or pink mycelium may develop on the surface.

spongy in texture, and tend to form hollow cavities that often have a fluffy appearance from growth of the fungus. If dry rot lesions are invaded by bacteria, they become slimy and foul smelling. Pockets of rotting tubers frequently develop around dry-rotted tubers invaded by bacterial soft rot. Dry rot takes several months to develop fully.

Fusarium sambucinum and *Fusarium solani* are the primary soil fungi involved in dry rot. They are present in all soils and survive for many years as resistant spores; infections are caused by spores carried in soil on tuber surfaces. *Fusarium* cannot penetrate intact tuber skin, lenticels, or suberized seed pieces; therefore, wounds are required for infection. Scab or blight lesions, wounds caused by insects or nematodes, bruises, and cuts are common entry sites. In the presence of a tuber wound, fungus spores germinate and enter the tuber, where the fungus slowly decays the tissue, causing typical dry rot symptoms unless bacterial soft rot develops. Damp conditions in storage favor soft rot development.

Proper handling and curing can provide economic control of dry rot in storage. Allow tubers to mature before harvest and prevent bruising or wounding of tubers during harvest. Wound healing reduces infection by *Fusarium*, so for the first 2 to 3 weeks of storage, hold tubers at 55°F (12.8°C) with good ventilation and a relative humidity of at least 95%. Fungicide treatments before storage may be useful when tubers must be harvested under adverse conditions or when conditions are not favorable for adequate wound healing.

Seed Piece Decay
Fusarium spp. and *Pectobacterium carotovorum* (*Erwinia carotovora*)

Seed piece decay occurs wherever potatoes are grown. It is most severe when unfavorable conditions occur after planting. Seed piece decay is managed with healthy seed, proper seed handling and planting procedures, and fungicide seed-treatments when necessary.

The first signs of seed piece decay are reddish brown to black spots that slowly form depressions on the surfaces of cut seed pieces. Later these spots may become black and slimy from bacterial infections. Entire seed pieces will rot if enough infections develop. Seed piece decay reduces stands, results in variable plant size, and may increase the incidence of blackleg or bacterial early dying.

Seed piece decay is usually caused by the combined action of the same *Fusarium* fungi that cause dry rot and the same *Pectobacterium* (*Erwinia*) bacteria that cause blackleg, although decay can sometimes be caused by *Fusarium* alone. *Fusarium* spores are present virtually everywhere in the soil, in storage, and on handling equipment. They germinate and infect the freshly cut surfaces of seed pieces, causing lesions that later may be invaded by *Pectobacterium*. When seed pieces are infected by both *Pectobacterium* and *Fusarium*, losses from seed piece decay are much greater than when *Fusarium* is present by itself. Cool soil that is too wet or too dry favors the development of seed piece decay.

Seed piece decay can be controlled adequately when seed pieces are planted under conditions that favor rapid suberization; *Fusarium* cannot infect cut surfaces after they are suberized. Warm the seed tubers to 50°F (10.0°C) before cutting, and keep cutting and handling equipment disinfected. Plant when the soil temperature is at least 45°F (7.2°C) and when soil moisture is 60% to 80% of field capacity. Make sure that seed pulp temperatures are higher than soil temperatures

GARY D. KLEINSCHMIDT

Seed piece decay caused by *Fusarium* appears as reddish brown lesions that may cover the entire seed piece. Decayed seed pieces may turn black and slimy if invaded by bacteria.

at planting. If possible, avoid irrigating before emergence. When planting conditions are likely to favor seed piece decay, treat cut seed pieces with a registered fungicide. Strains of *Fusarium* resistant to seed treatments containing benzimidazole fungicides have been reported.

Pink Rot
Phytophthora erythroseptica

Pink rot is a tuber decay that occurs sporadically in many of the growing areas in the West. It is important in parts of the Klamath Basin and Columbia Basin growing areas, where it can be a serious tuber decay problem. The causal organism, the water mold *Phytophthora erythroseptica*, causes disease on potatoes, tulips, and certain woody vines and ornamentals. The pathogen survives for long periods in many soils and becomes active when the soil is nearly or completely saturated with water. The disease is much worse when saturated soil is accompanied by warm temperatures. The only control is to avoid poorly drained soils and saturated soil conditions, especially at or near harvest.

Phytophthora erythroseptica infects roots and stolons as well as tubers, and it sometimes causes a wilting and early dying of plants that may be confused with Verticillium wilt. The most important phase of the disease is tuber decay. Infection spreads from eyes or lenticels through the tuber. When the infected tuber is cut, the rotted portion can be delineated by a dark line at its margin or a subtle color change. The decayed tissue is spongy but not discolored. With exposure to air, the surface of the decay turns a salmon pink color in about half an hour, turning to brown and then black after about an hour.

Tuber decay can spread in storage when liquid oozes from decaying tubers onto neighboring tubers, creating conditions favoring bacterial soft rot. To prevent rot in storage, avoid harvesting wet tubers whenever possible, maintain good airflow, avoid the accumulation of moisture on tubers, and keep the temperature as low as possible. The pathogen is inactive below 40°F (4.4°C).

Applying an effective fungicide to the crop during tuber initiation can reduce losses to pink mold. Resistance to commonly used fungicides containing metalaxyl has been reported in some areas. Check with local authorities for the latest recommendations.

Leak (Water Rot)
Pythium spp.

The water mold *Pythium ultimum* causes a tuber decay called leak, watery wound rot, or shell rot in most potato-growing areas; *P. aphanidermatum* causes water rot in Arizona and other warm-weather growing areas. The disease symptoms are similar to pink rot tuber decay, the causal organisms are similar, and subsequent control measures are essentially the same. In certain areas of Arizona, *P. aphanidermatum* can cause a severe stem rot similar to blackleg when fields are overirrigated late in the season. Total losses from tuber decay may occur if infested fields are irrigated after vine death and before harvest.

In the Columbia Basin, tuber decay caused by *P. ultimum* tends to occur on tubers from fields planted to potatoes for the first time, and disease levels are reduced or disappear after one or two crops of potatoes.

Pythium infests the tuber surface in the field, but it infects tubers only through wounds in the periderm. Infection may occur in the field or after harvest, and leak symptoms may develop in the field or in storage. The disease does not spread in storage. Seed tubers wounded during planting or that are planted without proper wound healing can be affected by leak.

Tuber tissue decaying with pink rot has a dark margin, but is not discolored when the tuber is first cut (center). The cut surface of a pink rot lesion turns salmon pink in about half an hour (left), then black after about an hour (right).

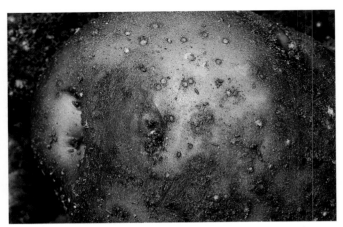

Tubers infected by *Pythium* develop lesions that appear dark on the tuber surface.

Lesions caused by *Pythium* have dark margins. Diseased tissue may appear pink at first, similar to pink rot lesions, then turns gray and eventually almost black. Affected tissue is soft and watery.

To control decay caused by *P. ultimum*, maintain adequate humidity and ventilation throughout the storage period. Allowing the tuber skin to mature before harvest reduces infections. Avoid harvesting during hot, dry weather because the pathogen is most active at 77° to 86°F (25° to 30°C). The incidence of leak may be reduced by applying effective fungicides during tuber initiation. Resistance to fungicides containing metalaxyl have been reported in some areas.

Stem Rot
Sclerotium rolfsii

The fungus *Sclerotium rolfsii* attacks many crops in warm climates, including alfalfa, tomatoes, carrots, beans, and potatoes. In the West, the disease causes significant potato losses in some areas of Kern County, California, rotting tubers in the field or in transit. Stem rot can be controlled with cultural practices or chemical treatment of heavily infested soil.

Sclerotium rolfsii survives in the soil as dark brown to black sclerotia the size of mustard seed that form on infected stems and tubers. Sclerotia germinate when the weather is warm, and the fungus invades dead or dying stem tissue at or below the soil surface, growing along stolons to tubers. *Sclerotium* is most active when the temperature is about 80° to 90°F (27° to 32°C), growing as a thin, white, cobwebby layer and penetrating into tubers through lenticels or from dead stolons. Stem rot appears near the end of the season, sometimes after vines are completely dead, and may destroy tubers before harvest. Because the pathogen grows rapidly, the entire process from invasion of stems through tuber decay can occur in just a few days. Infected tubers may decay in transit or in storage.

Damage from stem rot can be minimized by appropriate cultural practices. Maintain plants in a green condition for as long as possible prior to harvest. Harvest as soon as possible after irrigation has stopped and vines have been killed. If harvest is delayed, apply only as much water as absolutely necessary. Use fields known to be infested with *Sclerotium rolfsii* for early plantings to avoid having tubers in the ground when temperatures are most favorable for stem rot. Avoid moving soil from fields with a history of stem rot to noninfested areas.

Preplant treatment of infested soils with metam sodium will control stem rot for at least one season. Applying ammonium bicarbonate through sprinklers slows the growth of the pathogen but does not affect sclerotia. A single treatment delays disease development about 3 days, which can be important during the critical period between vine death and

Sclerotium rolfsii forms thin, white, cobwebby mycelium on stems and tubers at the end of the season.

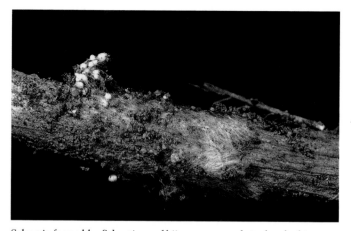

Sclerotia formed by *Sclerotium rolfsii* are poppyseed sized and white at first, turning dark brown to black.

harvest. However, if harvest is delayed and soil temperatures are high, ammonium bicarbonate will provide little or no benefit. Consult local authorities to get the latest recommendations for using chemical applications to control stem rot.

Powdery Scab
Spongospora subterranea

Powdery scab of potato is caused by a water mold, *Spongospora subterranea*, that occurs in most potato-producing areas of the world. Disease development is favored by cool, moist conditions, and symptoms are most severe on white and red-skinned tubers.

The powdery scab pathogen infects lenticels and wounds in tuber surfaces. Tuber cells are stimulated to divide and enlarge, causing the formation of small, purplish-brown, raised lesions. The pimple-like lesions or pustules are about ⅛ to ¼ inch (3 to 6 mm) in diameter. They gradually darken

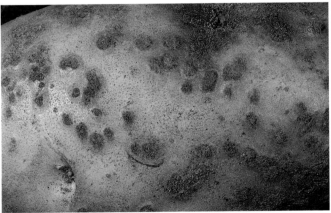

Powdery scab causes the formation of superficial pustules on the surface of potato tubers. These pustules release spores of the pathogen into the soil.

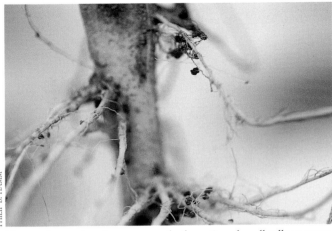

PHILIP B. HAMM

The powdery scab pathogen causes the formation of small galls on roots. The galls break down to release masses of pathogen spores.

and decay, leaving shallow depressions filled with powdery masses of dark brown spore balls. Roots may also be infected, resulting in the formation of small, white galls that appear similar to galls caused by root knot nematodes. These galls break down to release masses of spore balls.

Spongospora subterranea survives in the soil as resting spores originating from the spore balls released from tuber and root lesions. Stimulated by the presence of susceptible host roots, resting spores germinate to produce mobile spores called zoospores that swim through soil water to infect roots or stolons. These infections produce additional zoospores that spread infection to roots and tubers.

Resting spores survive in the soil for up to 6 years and can survive passage through the digestive tract of animals. The fungus can also infect and produce resting spores on the roots of other potato family plants such as tomato and nightshades. Powdery scab is much more severe on red- and white-skinned tubers. Tuber symptoms are less likely to occur on russets, but root galls are formed, so planting russet varieties will maintain soil populations of the pathogen at high levels.

Powdery scab can be minimized by using good cultural practices. Use pathogen-free seed tubers, do not use manure from farm animals that have fed on infected tubers, maintain optimum soil moisture, and avoid moving soil from contaminated fields to noninfested areas. Applications of metam sodium have been effective in reducing soilborne inoculum but may not be effective in controlling tuber infections. Consult local authorities for recommended application rates.

FOLIAR DISEASES

Important foliar diseases of potatoes include late blight, early blight, powdery mildew, and white mold. They are caused by pathogens that infect foliage, forming lesions on leaves or stems. Early and late blights can also infect tubers. Because the pathogens can be dispersed by wind, foliar diseases can spread rapidly, and timing of control measures usually is critical. Careful monitoring is required in areas where these diseases may develop.

Late Blight
Phytophthora infestans

Late blight is the most important disease of potatoes worldwide. Climatic conditions may limit the importance of late blight in the western United States, though newer, more aggressive strains of the pathogen now cause yearly epidemics in areas where conditions were once thought to be restrictive to late blight development. Late blight causes losses in potatoes as both foliar and tuber disease, and also affects other plants of the potato family such as tomatoes, eggplants, peppers, nightshades, and groundcherries. Late blight is controlled by eliminating cull piles and volunteer potatoes, using seedpiece fungicide treatments, properly applying foliar fungicides, and using proper harvesting and storage practices.

Symptoms and Damage

Lesions appear on leaves as small, water-soaked spots, usually at the tips or margins. Under the moderate temperatures and moist conditions that favor late blight development, lesions grow to large, brown to purplish black lesions with pale green halos of water-soaked tissue at their margins. In humid or wet weather, a white growth of mycelium and spores appears, usually on the underside of leaves. Lesions may kill entire leaflets and spread over the whole plant. Greasy, dark lesions frequently develop on stems, causing severe wilting or sometimes girdling and killing the plant. Bacterial stem rot often develops on late blight stem lesions. Late blight is most likely to occur in the wettest areas of the field, including low spots, the centers of center-pivot irrigation systems, and areas subject to poor air drainage.

Lesions appear on the tuber surface as irregular, brownish purple areas that have black "eyebrow pencil" marks at their margins. A coppery brown granular rot that lacks distinct margins extends a short distance into tuber tissue. If tuber lesions leak in storage, conditions favoring secondary bacterial invasion are created, and pockets of bacterial soft rot develop. Soft rot following late blight infections is responsible for major losses in storage.

Seasonal Development

The pathogen that causes late blight, the water mold *Phytophthora infestans*, overwinters in infected tubers in storage, fields, or cull piles. Plants originating from infected seed may produce lesions that extend above ground. The disease develops and spreads during periods of sprinkler irrigation, rainfall, or high humidity (above 90%), most rapidly when

Late blight lesions on leaves are brown to purplish black, with pale green margins. White mycelium and spores may develop.

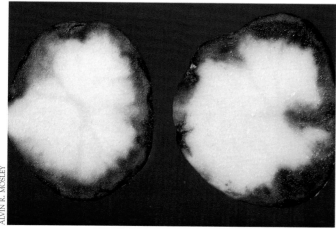

Late blight tuber lesions consist of brown, granular, rotted tissue that extends a short distance into the tuber.

Greasy, dark lesions frequently develop on stems of potato plants with late blight. These lesions may girdle and kill affected stems.

temperatures are moderate—68° to 77°F (20° to 25°C). When weather conditions are favorable, spores from infected foliage spread the disease to other plants. Infected seed tubers are usually the most important source of late blight inoculum. Infected tubers in cull piles also can be important, producing diseased volunteer plants that become an inoculum source for field infections. Late blight can spread rapidly. A single blighted leaf in an acre of potatoes can devastate the entire field if conditions are favorable for the disease. Tubers at or near the soil surface can be infected by spores washed into the soil from leaves or stems, or they can be infected by zoospores from sporangia produced on leaf and stem infections. Tubers may also be infected if they come in contact with spores during harvest.

Management Guidelines

Plant disease-free seed tubers, remove cull piles, and control potato volunteers to minimize sources of late blight. Use seed treatment fungicides to reduce spread of the pathogen and new infections during seed cutting operations.

Where late blight occurs every year, foliar fungicide applications are recommended during the season. Protectant fungicides must be applied before infections occur if they are to be effective. Follow these guidelines (the "Five Ps") for most effective use of fungicides.

- **Proper timing.** Begin applying fungicides just before row closure or according to the recommendation of a late blight prediction model for your area (see "Late Blight Forecasting," below).
- **Proper material.** Consult local authorities and the references listed under "Integrated Pest Management" and "Pesticide Application and Safety" in the suggested reading. Do not rely entirely on copper- or tin-based materials.
- **Proper rate.** Use the full rate recommended on the label for the material you are using.
- **Proper frequency.** Depending on the materials you are using, weather conditions, and the occurrence of late blight in your area, repeat applications every 5 to 14 days.
- **Proper application method.** Ground applications can be the most effective for applying higher levels of fungicide residues. Aerial application is less effective than ground application, but it can still be very effective if applied without skips, at the proper height, and under proper climatic conditions. Application through sprinklers (chemigation) is least effective for applying fungicide residues, hence the rapid loss of effective fungicide residue in overhead-irrigated fields. Alternating between aerial application and chemigation achieves better control than chemigation alone and reduces the cost of using only aerial. If using this combination, begin by using air, then follow with chemigation.

Where late blight occurs sporadically, apply fungicides as soon as lesions appear in the area and field conditions favorable for blight development are expected to continue for more than 2 or 3 days.

Leave tubers in the ground for at least 2 weeks after vine death to allow *Phytophthora* spores produced on the foliage to die before tubers are harvested, otherwise even a small amount of foliar blight can cause substantial tuber decay in storage. Avoid harvesting under wet conditions. If wet tubers go into storage, use increased ventilation and shut off humidifiers to remove free moisture from them as quickly as possible. Keep the storage temperature as low as allowed by the destined market and maintain proper ventilation and humidification to avoid free moisture in the piles.

Late Blight Forecasting. Successful blight forecasting models have been developed for scheduling fungicide applications to control late blight. These models use records of temperature, rainfall, relative humidity, and the previous season's disease history to forecast the development of late blight. Current predictions from these models are available online and via telephone hotlines during the season for several growing areas. The latest information on accessing these models can be found on the Potato Information Exchange Web site (http://oregonstate.edu/potatoes) under Pest Management, Diseases and Physiological Disorders.

Early Blight
Alternaria solani

Early blight can affect both foliage and tubers wherever potatoes are grown, reducing yield and tuber quality if not controlled. The disease is most serious in Rocky Mountain growing areas and parts of the Snake River Valley. Proper management of soil moisture and fertility reduces the severity of early blight. Foliar blight is controlled with properly timed fungicide applications, and tuber infections are reduced by delaying harvests. Early blight attacks tomatoes and other potato family plants, but the strain of pathogen that attacks tomato is different from the one that attacks potato.

Symptoms and Damage

Early blight is primarily a disease of stressed or aging (senescing) plants. Early blight infections are characterized by brown lesions that appear first on the oldest (lower) leaves. When small, the lesions are usually circular, about ⅛ to ⅜ inch (3 to 9 mm) in diameter, and consist of concentric rings of dead tissue that give a targetlike or bull's-eye appearance; hence, early blight is sometimes referred to as "target spot." In contrast to late blight, where developing lesions have a nearly rounded edge, early blight lesions become angular in shape because expansion is limited by leaf veins. A yellow zone that fades into the green of healthy tissue often surrounds the lesion. Affected leaves usually do not fall off.

As the disease begins to spread, lesions appear higher up on the younger leaves of the plant. Dark brown to black stem lesions may also occur at this time. Early blight symptoms are more severe on vines dying from natural aging or stressed by insufficient nutritional levels or other diseases such as Verticillium wilt. If early blight foliage infections become severe, harvest yields can be reduced significantly. Tuber infections that occur during harvest can result in significant storage losses and reduced tuber quality.

Tubers are infected at harvest by *Alternaria* spores deposited on the soil during the foliar phase of the disease. Mature tubers usually must be wounded during harvest for infection to occur, and free moisture must be present on the tuber surface. Therefore, tubers grown in coarse, sandy soil and tubers harvested when wet are most likely to be affected. Tuber lesions appear during storage, usually several months following harvest. Lesions are small and brown to black, with surface margins that are purple to black in color. Where they penetrate into the tuber flesh, they are brown and slightly leathery, and they usually remain small and superficial under good storage conditions. Tuber lesions reduce the market value of the crop. In parts of Idaho, Colorado, and other Rocky Mountain growing areas, tuber lesions are the most serious phase of the disease, particularly with the smooth, white-skinned cultivars commonly grown for chipping.

Seasonal Development

The early blight fungus, *Alternaria solani*, overwinters on potato refuse in the field, on tubers, and on other plants of the potato family. Spores produced by the fungus come in contact with leaves touching the soil or are carried to leaf surfaces by wind. In the presence of free moisture, spores germinate and the fungus penetrates directly through leaf surfaces. The fungus grows within the plant for a time without causing symptoms; the first lesions usually appear during early tuber growth, about 1 week after flowering. Symptoms are more pronounced on leaves that are weakened by age, stress, poor nutrition, or diseases such as Verticillium wilt. In some areas, early blight does not occur until late in the season.

Secondary spread occurs when spores produced on leaf lesions are carried to other sites by air movement. The most rapid spread of early blight occurs in alternating wet and dry weather. Moist foliage conditions created by heavy dews, frequent rains, sprinkler irrigation, or high humidity are necessary for spore germination and infection. However, dry conditions favor wind transport of spores from leaf lesions

GARY D. KLEINSCHMIDT

Early blight lesions on tubers appear sunken and are brown or black. They usually remain superficial.

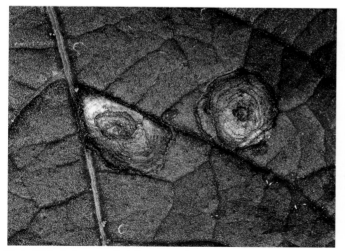

Early blight lesions are dark brown, with concentric rings that have a target-like appearance. They become angular in shape when limited by leaf veins.

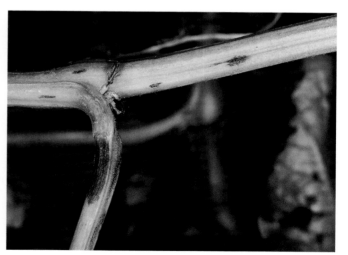

Dark brown early blight lesions develop on stems in later stages of the disease.

to other foliage. Appearance of early blight lesions is an indication that secondary spread is occurring. Spore trapping may detect secondary spread of *Alternaria solani*. Spore traps generally detect spread before lesion development is readily apparent in the field.

Tubers become infected as they are lifted through soil infested with *Alternaria* spores, which tend to be concentrated at the soil surface. Conditions that increase the severity of foliar blight increase the quantity of spores on the soil surface at harvest. Because tuber infections occur mostly through wounds, immature tubers and tubers of white- and red-skinned cultivars are infected more readily. Harvesting tubers from coarse, sandy soils can also increase disease severity by increasing injury. Letting tubers mature after vine death allows proper skin development and decreases the amount of tuber injury and tuber infection. Harvesting when vines are green increases tuber infections because tubers are immature. Infections can also occur through lenticels, which are open when wet; therefore, tuber infections are higher on tubers grown under wet conditions and on tubers harvested when wet.

Management Guidelines

In many growing areas, appropriate cultural practices are usually sufficient to avoid economic loss from early blight. Practices that optimize growing conditions through proper fertilizer, water, and pest management reduce stress and therefore decrease the severity of early blight. Generally, late-maturing cultivars are less susceptible to the foliar phase of the disease. However, if tubers are not allowed to mature properly before harvest, tuber infection can still be severe.

Nitrogen and phosphorus deficiencies increase susceptibility to early blight. Use petiole analysis as a guide to maintain levels of nitrogen and phosphorus considered adequate for your growing area. Make sure potassium levels are adequate; otherwise the tuber skin will not develop properly and tuber infection may increase.

Reduce early blight infections of tubers by letting tubers mature in the ground for at least 2 weeks after vine death. Avoid harvesting wet tubers. If conditions force you to harvest wet tubers, dry them as quickly as possible with forced-air ventilation as soon as they are placed in storage. Follow procedures that minimize mechanical injury during harvesting and handling. In all situations, conditions that favor rapid wound healing during the first 2 weeks of storage will reduce the severity of tuber infection.

Application of foliar fungicides is economically justified in areas where the disease becomes severe before vine death begins from other causes or where tuber blight is a problem. Early blight controls are presently recommended in some of the Columbia Basin, Klamath Basin, Snake River Valley, and Rocky Mountain growing areas. To be effective, applications must be properly timed and the fungicides must thoroughly cover the vines. Foliar sprays should be applied when secondary spread of early blight begins. Degree-day models, spore trapping, and field monitoring may be used to determine when secondary spread begins and spray programs should be started. Guidelines used in Colorado and Idaho for timing early blight sprays are given below. Sprays for early blight control have a greater effect on tuber yield when other diseases that can mask the beneficial effects of early blight control, such as Verticillium wilt, are controlled. Fungicide applications are not always economically justified, because tuber blight may not be a problem and foliar blight may be mild or occur too late in the season to cause significant loss. Consult your local extension agent, farm advisor, or other experts for specific recommendations concerning your area.

Using Degree-days in Colorado. A degree-day model has been developed for scheduling early blight sprays in Colorado, where early blight is a problem every year if not controlled. The model estimates the time of first lesion appearance and therefore secondary spread of early blight by using heat units calculated as degree-days. The time of first lesion appearance depends on the amount of heat the plant and pathogen experience, so measuring the amount of heat accumulating over time is biologically more meaningful and more accurate than using calendar days to predict outbreaks. A degree-day (DD) is the amount of heat that accumulates during a 24-hour period when the average temperature is 1 degree above the developmental threshold for the pest or

Date	Daily low	Dailiy high	Degree-days for date	Total degree-days for season to date
5/17	35.°F	53.°F	0	0
5/18	25	59	0	0
5/19	31	68	4.5	4.5
5/20	36	70	8.0	12.5
5/21	35	71	8.0	20.5
5/22	37	73	10.0	30.5
5/23	37	68	7.5	38.0
5/24	35	66	5.5	43.5
5/25	22	61	0	43.5
5/26	25	67	1.0	44.5

Figure 34. Sample form for recording degree-days in Colorado. Record the daily high and low temperatures and the degree-days for the date. Use Table 14 or calculate the degree-days as average daily temperature (in °F) minus 45, recording negative numbers as zeroes. The total degree-days for the season to date will tell you when to schedule the first spray application for early blight.

plant in question. The Colorado early blight model accumulates degree-days as degrees Fahrenheit and uses a developmental threshold of 45°F (7.2°C).

To use degree-days for scheduling treatments, start recording daily minimum and maximum temperatures in your field after the earliest planting in your area. Use Table 14 to determine the total degree-days for each day. After planting, begin to tally degree-days for the season using a form such as the one in Figure 34. To properly calibrate the model for a given production area, simultaneously observe lower leaves for the time of first lesion appearance. Begin looking for lesions before the first flowers form. Spore-trapping data can also help determine when secondary spread is occurring.

In the San Luis Valley, apply the first early blight spray when the degree-day total reaches 650 DD; in northeastern Colorado, make the first blight application when the degree-day total reaches 1,125 to 1,150 DD. The model is used only to time the initial application properly. Make subsequent applications according to label directions. Degree-day models such as this may be used for other areas where early blight requires control, but they must be calibrated for several years in each area to ensure their accuracy.

Spore Trapping. Trapping and identification of *Alternaria solani* spores can be used to determine when early blight is beginning to spread within an area. Spores are usually detected with spore traps before early blight is readily detected in the field. To use spore trapping, mount greased microscope slides in a weather-vane spore trap above the level of the potato canopy in a potato field. A weather station is a convenient location. Begin trapping about the time tuber initiation begins. Replace the slides with fresh ones at least twice a week and examine the exposed slides with a microscope to see if *A. solani* spores are present. The spores of *Alternaria* are easily recognized, but in some locations *A. alternata* spores may also be trapped. The two species can be distinguished by the spore shape and pattern of cross-walls, or septa, within each spore (Fig. 35). The presence of spores indicates that secondary spread of the pathogen is occurring. In areas where early blight is a perennial problem and a degree-day model is not available, you may choose to begin fungicide applications when spores appear. In other areas, begin monitoring fields for early blight symptoms at this time.

Spore trapping data may be used to develop a degree-day model. If the appearance of *A. solani* spores occurs consistently from season to season in a given area once a certain number of degree-days have accumulated, that degree-day total indicates when to make early blight management decisions.

Field Monitoring, Idaho Guidelines. In the upper Snake River Valley growing areas of Idaho, apply fungicides at the first sign of secondary spread, as indicated by the appearance of target spot lesions in the upper two-thirds of the plant.

Table 14. Degree-Days Above 45°F (7.2°C), Based on Daily Maximum and Minimum Temperatures. To find the degree-days for a day, follow the column and row of the day's minimum and maximum temperatures to where they intersect. For odd-numbered temperatures, interpolate between the table values.

MINIMUM TEMPERATURES

MAX TEMPS	70	68	66	64	62	60	58	56	54	52	50	48	46	44	42	40	38	36	34	32	30	28	26	24	22	20
120	50	49	48	47	46	45	44	43	42	41	40	39	38	37	36	36	35	34	34	33	33	32	32	31	31	30
118	49	48	47	46	45	44	43	42	41	40	39	38	37	36	35	35	34	33	33	32	32	31	31	30	30	30
116	48	47	46	45	44	43	42	41	40	39	38	37	36	35	34	34	33	32	32	31	31	30	30	29	29	29
114	47	46	45	44	43	42	41	40	39	38	37	36	35	34	33	33	32	31	31	30	30	29	29	28	28	28
112	46	45	44	43	42	41	40	39	38	37	36	35	34	33	32	32	31	30	30	29	28	28	27	27	27	27
110	45	44	43	42	41	40	39	38	37	36	35	34	33	32	31	31	30	29	29	28	28	27	27	27	26	26
108	44	43	42	41	40	39	38	37	36	35	34	33	32	31	30	30	29	28	28	27	27	26	26	26	25	25
106	43	42	41	40	39	38	37	36	35	34	33	32	31	30	29	29	28	27	27	26	26	25	25	25	24	24
104	42	41	40	39	38	37	36	35	34	33	32	31	30	29	28	28	27	26	26	25	25	24	24	24	23	23
102	41	40	39	38	37	36	35	34	33	32	31	30	29	28	27	27	26	25	25	24	24	24	23	23	22	22
100	40	39	38	37	36	35	34	33	32	31	30	29	28	27	26	26	25	24	24	23	23	23	22	22	21	21
98	39	38	37	36	35	34	33	32	31	30	29	28	27	26	25	25	24	23	23	23	22	22	21	21	21	20
96	38	37	36	35	34	33	32	31	30	29	28	27	26	25	24	24	23	23	22	22	21	21	20	20	20	19
94	37	36	35	34	33	32	31	30	29	28	27	26	25	24	23	23	22	22	21	21	20	20	19	19	19	18
92	36	35	34	33	32	31	30	29	28	27	26	25	24	23	22	22	21	21	20	20	19	19	18	18	18	17
90	35	34	33	32	31	30	29	28	27	26	25	24	23	22	21	21	20	20	19	19	18	18	18	17	17	17
88	34	33	32	31	30	29	28	27	26	25	24	23	22	21	20	20	19	19	18	18	17	17	17	16	16	16
86	33	32	31	30	29	28	27	26	25	24	23	22	21	20	19	19	18	18	17	17	16	16	16	15	15	15
84	32	31	30	29	28	27	26	25	24	23	22	21	20	19	18	18	17	17	16	16	15	15	15	14	14	14
82	31	30	29	28	27	26	25	24	23	22	21	20	19	18	17	17	16	16	15	15	15	14	14	14	13	13
80	30	29	28	27	26	25	24	23	22	21	20	19	18	17	16	16	15	15	14	14	13	13	13	12	12	12
78	29	28	27	26	25	24	23	22	21	20	19	18	17	16	15	15	14	14	13	13	12	12	12	11	11	11
76	28	27	26	25	24	23	22	21	20	19	18	17	16	15	14	14	13	13	12	12	12	11	11	11	11	10
74	27	26	25	24	23	22	21	20	19	18	17	16	15	14	13	13	12	12	11	11	11	10	10	10	10	10
72	26	25	24	23	22	21	20	19	18	17	16	15	14	13	12	12	11	11	10	10	10	9	9	9	9	9
70	25	24	23	22	21	20	19	18	17	16	15	14	13	12	11	11	10	10	9	9	9	9	8	8	8	8
68		23	22	21	20	19	18	17	16	15	14	13	12	11	10	10	9	9	9	8	8	8	8	7	7	7
66			21	20	19	18	17	16	15	14	13	12	11	10	9	9	9	8	8	7	7	7	7	7	6	6
64				19	18	17	16	15	14	13	12	11	10	9	8	8	8	7	7	7	6	6	6	6	6	6
62					17	16	15	14	13	12	11	10	9	8	7	7	6	6	6	5	5	5	5	5	5	5
60						15	14	13	12	11	10	9	8	7	7	6	5	5	5	5	5	4	4	4	4	4
58							13	12	11	10	9	8	7	6	6	5	5	5	4	4	4	4	4	3	3	3
56								11	10	9	8	7	6	5	5	5	4	4	4	3	3	3	3	3	3	3
54									9	8	7	6	5	4	4	3	3	3	3	2	2	2	2	2	2	2
52										7	6	5	4	3	3	2	2	2	2	2	2	2	2	1	1	1
50											5	4	3	2	2	2	1	1	1	1	1	1	1	1	1	1
48												3	2	1	1	1	1	1	1	1	1	0	0	0	0	0

Figure 35. The large, dark, segmented spores of *Alternaria* are easy to identify with a compound microscope. Spores of *A. solani* (left) can be distinguished from those of *A. alternata* (right) by a long "beak" that is usually the same length as or longer than the spore body. The beak of *A. alternata* spores is not more than one-third the length of the spore body.

Careful monitoring is required. The number of sprays recommended depends on the time between the beginning of secondary spread and vinekill. These recommendations may be adaptable to other areas where early blight may cause significant losses, such as the Klamath Basin and other California and Rocky Mountain growing areas.

- Do not spray until blight lesions appear on leaflets above the lower one-third of the plant.
- If the first spray is near July 9, follow with three more sprays 10 to 12 days apart.
- If the first spray is near July 21, follow with two more sprays 10 to 12 days apart.
- If the first spray is near August 1, follow with one more spray 10 to 12 days later.
- If the first spray is near August 10, check closely to see if another spray is needed.
- If early blight builds up after August 15, do not spray in eastern Idaho; one spray might be applied in south-central Idaho, depending on harvest date.

White Mold
Sclerotinia sclerotiorum

The fungus *Sclerotinia sclerotiorum* infects a large number of crop and wild plant species. In potato-growing areas it commonly causes disease on beans, lettuce, carrots, and alfalfa. Its occurrence on potatoes has increased with the use of sprinkler irrigation. White mold on potatoes is seen regularly in areas of Idaho, Washington, Oregon, and California, and is controlled by managing irrigation, avoiding excess nitrogen fertilization, and applying fungicides.

Sclerotinia overwinters as hard, black, irregularly shaped sclerotia that are formed inside the infected stems of potatoes and other hosts. When exposed to moisture for prolonged periods above 60°F (15.6°C), sclerotia germinate and grow into light brown, funnel-shaped or flat fruiting structures called apothecia, which are ¼ to ⅜ inch (6 to 9 mm) in diameter. These structures are formed over a period of about 6 weeks and eject spores that infect nearby plants or can be carried for miles on wind currents. Most spore dispersal occurs within the potato field. The spores germinate and infect senescing leaves and blossoms when plant surfaces are wet for an extended period of time. Infected petals that fall and stick to lower stems or infected senescing leaves become the avenue allowing the pathogen to spread into stems. Water-soaked lesions and white mold then develop. Stems may be girdled when infections are severe. Tubers near the soil surface can be infected, but this phase of the disease is seldom seen in the western states.

White mold is associated with conditions favoring prolonged presence of water on stems and leaves. It is more prevalent in the wettest parts of fields and tends to occur after the vines close between rows, when free moisture remains longer on stems and leaves.

Avoid excess nitrogen, which creates heavy canopy growth that promotes conditions favorable for white mold.

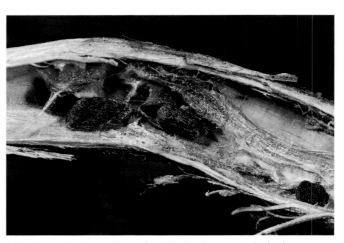

Sclerotinia sclerotiorum forms hard black sclerotia inside dead stems.

Light brown lesions and water-soaked areas with white, fluffy mycelium develop on stems infected by *Sclerotinia*. A black sclerotium can be seen forming on the surface of this lesion.

The earliest symptoms of powdery mildew are small, light brown stipples on leaflets, stems, and petioles.

Watch for the appearance of apothecia or disease symptoms after plants have emerged and as vines close. If white mold infections appear, reduce irrigation to allow lower parts of the plants to dry more quickly. After vine closure, apply water less often to make sure that lower leaves and stems do not remain continuously wet for long periods. Reduce irrigation rates if white mold appears on fallen leaves. Be sure changes in irrigation schedules do not allow soil moisture to drop below adequate levels. White mold can be reduced with fungicide applications. Most reports suggest that low or moderate levels of white mold have little impact on yield even when symptoms are readily apparent in the field.

Powdery Mildew
Golovinomyces cichoracearum
(Erysiphe cichoracearum)

Powdery mildew is known to occur on potatoes in the Snake River Valley, the Columbia Basin, southern California, and in Utah. While the disease occurs under all irrigation situations, it requires control only in furrow-irrigated fields in some areas of the Columbia Basin and lower Snake River Valley.

Powdery mildew first appears as very small, elongated, light brown stipples on stems and petioles. These coalesce as they develop to form dark, water-soaked lesions. Symptoms ultimately spread to leaves. Sometimes the spore-forming structures of the fungus develop in leaf lesions, giving leaves a powdery dusty brown or gray color that looks like soil or spray residue. As the disease progresses, lower leaves turn yellow and fall off. The plant stays erect but becomes covered with powdery mildew.

Sulfur will control powdery mildew if it is applied before infections are established. However, because symptoms appear after infections are established, sulfur applications made after

When spore-forming structures develop, powdery mildew lesions appear dusty brown or gray and powdery.

this time are ineffective. In Columbia Basin furrow-irrigated fields where the disease is expected, first sulfur applications are usually made in mid to late July and repeated every 2 weeks. Consult your local extension agent, farm advisor, or other experts for specific recommendations in your area.

VIRUS AND PHYTOPLASMA DISEASES

The causal agents of virus and viruslike diseases are viruses and phytoplasmas. Viruses cannot reproduce outside a host organism and are transmitted mechanically or by vectors such as aphids, leafhoppers, or nematodes. Phytoplasmas are bacterialike microorganisms that are transmitted by leafhoppers that feed on potato plants. Certified seed programs provide the most effective means for controlling these diseases, because in most instances the primary source of inoculum is infected seed tubers. Insecticides may control the primary vectors of the viruses and phytoplasmas that affect potatoes, and in most cases they reduce the spread of these diseases.

Leafroll

Potato Leafroll Virus (PLRV)

In the past, leafroll has caused severe losses in yield and quality of stored tubers, especially of Russet Burbank. Aphid monitoring programs and control of aphid weed hosts, treatment of aphid overwintering hosts, elimination of potato volunteers, use of certified seed tubers, and increased use of systemic insecticides have reduced the incidence of leafroll in recent years. However, the disease can still be a problem when control measures are ignored or when large aphid flights occur.

Symptoms and Damage

The greatest losses to leafroll are from rejection of seed lots and devaluation of commercial tubers with a speckling or netting of discolored vascular tissue called "net necrosis." Yield losses due to PLRV infection may also be significant. Russet Burbank is the main cultivar that develops net necrosis, while some cultivars do not express this symptom under any condition. Leafroll symptoms vary depending on whether infection is from current-season transmission or from tuberborne virus.

Current-season infections occur when plants are infected with PLRV by aphid vectors. Symptoms develop first on young leaves at the tops of plants because the virus tends to be active only in rapidly growing tissue. Young leaflets stand upright, roll upward, and turn pale or yellow. The rolling usually starts at the base of the infected leaflet. A pink or reddish color may develop on the leaves, starting at the margins. Maximum symptom expression occurs at 65° to 75°F

(18° to 24°C). When primary infections occur late in the season, after new vine growth has stopped, foliar symptoms may not develop. Symptoms caused by blackleg bacteria and phytoplasmas can be confused with current-season leafroll symptoms. Injury from drift of herbicides or pesticides may

Upper leaves of plants with current-season PLRV infections stand upright, roll upward, and turn pale.

A pink or red color may develop on the leaflet margins of potato plants with current-season PLRV infections.

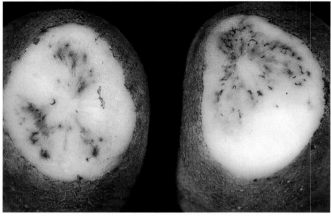

Brown discoloration of tuber vascular tissue, called net necrosis, may develop in most of the tubers from a plant infected with PLRV.

also be confused with PLRV symptoms. Identification of PLRV using serological techniques is the best means of diagnosis. Current-season infections can cause yield reductions in commercial potatoes, and net necrosis can cause substantial quality problems.

In certain varieties, including Russet Burbank and Ranger Russet, the food-conducting tissue (phloem) in stems, petioles, and tubers may develop a discoloration in response to current-season PLRV infection. In tubers, a translucent to dark brown speckling or netting (net necrosis) develops and may appear in the field or in storage. Not all tubers from an infected plant of a susceptible cultivar develop net necrosis. The actual number depends on the timing of PLRV infection and is greatest when infection occurs at the time of most rapid tuber expansion.

Other diseases or disorders cause tuber symptoms that may be confused with net necrosis. These include aster yellows, internal heat necrosis, internal brown spot, and discolorations of stem end vascular tissue caused by Verticillium wilt, or vinekill under certain conditions. When PLRV is suspected, an experienced plant pathologist should be consulted to assure proper diagnosis.

Chronic or seedborne infections occur when plants grow from seed tubers infected with PLRV. Symptoms of chronic infections develop on foliage throughout the plant. Leaflets roll up, turn slightly pale, become stiff, dry, and leathery, and make a papery sound when rubbed. Leaf rolling usually starts at the tips of the leaflets. Plants are often stunted or rigid with leaf yellowing, and the older leaflets may become discolored or turn brown and die. Chronic infections are much more severe than current-season infections, and are therefore responsible for greater yield reductions. However, chronic infections do not cause net necrosis in tubers.

Plants with chronic (tuberborne) PLRV infections are stunted, and symptoms develop on the lower leaves.

Seasonal Development

PLRV is carried in infected tubers and is transmitted from plant to plant by specific aphid vectors. All the aphid vectors have winged and wingless stages in their life cycle. The green peach aphid, *Myzus persicae*, reportedly transmits PLRV most efficiently. This aphid acquires the virus when feeding on infected host plants and can infect other plants after a latent period of about 24 hours or longer. Once infected with PLRV the aphid usually is able to transmit the virus for the rest of its life. Aphids are attracted to yellow infected plants and pick up the virus by feeding on these plants. Winged aphids are responsible for introducing the virus into a field and for long-distance spread. The virus can also be transmitted by wingless forms to adjacent plants, usually within a row or between rows. Large areas within a field can be infected and show leafroll symptoms due to spread by wingless aphids.

Most potato family plants and several mustard family weeds can be hosts for PLRV. However, they do not play a significant role as reservoirs of this virus in temperate regions. The only important virus sources are potato plants grown from infected seed tubers, volunteers that have grown from infected tubers left in the field or in cull piles, and other potato fields.

PLRV is introduced into the phloem, or food conducting tissue, of plants when virus-infected aphids feed. It moves through the phloem to growing leaves and expanding tubers; the more rapidly tubers are expanding at the time of infection, the more likely they are to be infected by the virus. If plants are infected when foliage growth has stopped but tubers are expanding, PLRV will infect the tubers without causing foliar symptoms, and net necrosis symptoms can still develop in susceptible cultivars such as Russet Burbank.

If a cultivar is susceptible to net necrosis, most of the tubers infected with PLRV will develop this symptom. It develops first in the stem end and spreads further into the tuber in the field or in storage. Infected tubers may be symptomless at harvest and still develop symptoms during storage.

Management Guidelines

The principal measures for controlling PLRV are the use of certified, PLRV-free seed tubers, insecticides to control aphid vectors, and control of potato volunteers. Some varieties are less susceptible to infection by the virus. Most varieties are resistant to the development of net necrosis, while Russet Burbank is the most susceptible (see Table 6 in the third chapter, "Managing Pests in Potatoes").

Grow seed potatoes in areas where sources of PLRV are limited and where aphid vectors do not appear until late in the season; use appropriate insecticide treatments when needed. Favored seed production areas in western states are typically high in elevation and well isolated from other potatoes and from peach, apricot, and plum trees. If early-season leafroll symptoms appear, remove each plant with symptoms, including any tubers present, and adjacent plants according to the diagram shown in Figure 31 (in the previous chapter). Kill vines as early as possible to minimize exposure to vector populations.

Minimize leafroll in commercial fields by planting certified seed, following recommended insecticide programs for controlling aphid vectors (see discussion of green peach aphid in the previous chapter), and controlling volunteers. Vector control is important throughout the season, but late-season control is especially important where cultivars susceptible to net necrosis are grown.

When certified, PLRV-free seed is used, volunteers become the major source of PLRV, and they therefore need to be controlled. Always leave the field as clean as possible after harvest. In winter growing areas, till the field immediately after harvest to expose tubers to as much heat and desiccation as possible. In areas with cold winters, leave tubers exposed. If excessive numbers of tubers have been left in the field, you may want to use rototilling or some other type of additional tillage to destroy as many of them as possible. Do not leave cull piles in or near the field or around storage or packing facilities. Whenever possible, destroy volunteers that appear in or near potato fields, as well as weeds such as nightshades, groundcherries, and mustard family weeds, which host the virus or aphid vectors. Do not plant back-to-back crops of potatoes. Use appropriate herbicides or cultivation to control volunteers that appear in rotation crops or field borders. Application of a suitable sprout inhibitor to vines before harvest can reduce emergence of volunteers the following year. An in-field treatment with sprout inhibitor may be worthwhile where the chance of leafroll infection is high because of a large green peach aphid population or the presence of a large PLRV reservoir.

Because net necrosis symptoms will intensify in storage, market tubers from fields with leafroll symptoms early and save tubers from fields without leafroll symptoms for prolonged storage. Storage at lower temperatures retards but does not prevent the development of net necrosis.

Mosaic-Type Potato Viruses

The mosaic-type potato viruses occur wherever potatoes are grown. This group of viruses includes potato viruses A (PVA), S (PVS), X (PVX), and Y (PVY), all of which are of concern in the western states. They are called latent viruses because they can be carried in plants or tubers and may reduce yields without causing easily seen foliar symptoms. Each virus has a number of different types or strains, and the symptoms caused by these strains may differ. Potato cultivar and weather conditions also affect the symptom expression, with the most pronounced symptoms appearing in cool, cloudy weather. Symptoms caused by chronic (tuberborne) infections are often more severe than those caused by cur-

Leaflets of plants infected with mosaic virus (middle and right) may be mottled with areas lighter green than normal and areas darker green than normal. Leaflets become wrinkled because the darker areas grow faster. Healthy leaves are shown on the left.

rent-season infections with the same virus strain. The presence of more than one of the viruses in a plant usually affects the types of symptoms and increases symptom severity. The more severe symptoms caused by mixed infections (e.g., PVX and PVY) are referred to as "rugose mosaic." Symptoms caused by different viruses can be similar, so the type of virus usually cannot be identified by symptoms alone. Field diagnosis is often limited to "mosaic virus." Positive identification requires the use of serological techniques or polymerase chain reaction (PCR).

All of these viruses can be carried in tubers that show no symptoms. Some of them can be transmitted mechanically by seed cutting or when a cut or open surface of an uninfected plant contacts the sap of an infected plant. Mechanical spread occurs more readily between plants when they are wet and can be caused by cultivation or other movement of equipment, people, or animals through potato fields. PVA and PVY are also transmitted by aphids. The most efficient aphid vector is the green peach aphid, but any aphid that probes or feeds in a potato field can transmit these viruses. Therefore, high numbers of aphids leaving nearby ripening grain or recently cut hay or other green crops can become significant vectors. Aphids can transmit either of these viruses immediately after feeding on an infected plant, but they lose the virus after they feed on one to three uninfected plants. These nonpersistent viruses (as opposed to the persistent, systemically carried PLRV) are usually transmitted only short distances. Because aphids are infective for only short times and preventing aphid movement into fields is very difficult, spread of PVA and PVY is difficult to control, and nearly impossible where high aphid numbers are common. Control of aphids with insecticides is not effective in limiting the spread of these nonpersistent viruses. PVS is usually not spread by aphids, and PVX is transmitted only mechanically.

The major control for potato viruses is the use of virus-free seed tubers. Special procedures are used to obtain virus-free plants and to identify the presence of these viruses, including PLRV, in tubers to be used for seed production. Steps must be taken to prevent virus infection during the production of certified seed tubers (see "Seed Quality and Seed Certification" in the third chapter, "Managing Pests of Potatoes"). In all potato production areas, disinfect seed cutting equipment with recommended disinfectants (see Table 7 in the third chapter), avoid cutting seed tubers that have sprouts large enough to be broken, and control potato volunteers and weeds, especially nightshades, which can be hosts for potato viruses.

Potato virus A (PVA)

The disease caused by PVA is known as mild mosaic, crinkle mosaic, and veinal mosaic. Leaves on affected plants are mottled, with some areas light green to yellow and some darker green than normal. Mottled areas may vary in size and occur both on and between leaf veins. Margins of affected leaves may be wavy, and leaves may appear slightly rugose where veins are sunken and interveinal areas raised. Rugose symptoms can also be caused by certain strains and combinations of PVX and PVY. Affected plants tend to open up because the stems bend outward. Symptoms are more pronounced and yield reductions are greater when PVX is also present. PVA is transmitted in a nonpersistent manner by at

least seven different aphid species, including the green peach aphid, *Myzus persicae*, and the potato aphid, *Macrosiphum euphorbiae*.

Potato virus S (PVS)

Although in most cultivars PVS is symptomless, some potato cultivars infected with some strains of the virus demonstrate slight deepening of leaf veins, and possibly stunting, mottling, or bronzing. There is argument about whether this virus causes any reductions in yield by itself, although losses of 10% to 20% have been reported. Combined presence of PVS and PVX reduces yields more than either virus alone.

Mature plants are resistant to the virus; plants must be infected early in the season for tubers to become infected. The disease is carried in tubers and is mechanically transmitted. PVS can be reduced by using certified seed tubers, but it is almost impossible to fully control in commercial fields. Special laboratory or greenhouse techniques must be used to identify the presence of PVS in seed stock because it is almost always latent.

Potato virus X (PVX)

The most widespread of the potato viruses, PVX is also called potato latent virus, potato mottle virus, and latent mosaic. Some strains of the virus produce no visible symptoms when no other viruses are present, although symptomless plants may have yields reduced 15% or more when compared to virus-free plants. Other strains of PVX may cause mild mottling, sometimes called weather mottling, under prolonged periods of low light intensity. Low tem-

Rugose mosaic symptoms (left), where veins appear sunken and interveinal areas raised, occur on leaves of plants infected by both PVX and PVY. Certain strains of PVX or PVY alone may cause rugose mosaic. Healthy leaves are shown on the right.

perature (60° to 68°F [15° to 20°C]) enhances symptoms, and the additional presence of PVA or PVY may cause crinkling, rugose mosaic, or browning of leaf tissue. Some combinations of virus strain and potato cultivar may result in the death of all or part of the plant and tuber necrosis.

PVX is carried in tubers and can be transmitted mechanically by machinery, spray equipment, root-to-root contact, sprout-to-sprout contact, or seed cutting equipment. There are no known aphid vectors. The spread of PVX is controlled by using certified seed tubers and avoiding mechanical transmission.

Potato virus Y (PVY)

The most severe of the mosaic-type potato viruses, PVY causes a range of symptoms from mild mosaic to defoliation and death, depending on the virus strain, potato cultivar, and environmental conditions. The disease caused by PVY is sometimes called potato vein banding or leaf drop. Yield losses to PVY are greater in Columbia Basin growing areas, presumably because of the longer growing season and normally higher yields. In contrast to the other mosaic viruses, some strains of PVY can also cause external and internal necrotic areas in tubers.

Two types of foliar symptoms are characteristic of PVY. Vein banding is the development of brown streaks along veins on the undersides of leaves. If vein banding becomes severe, leaf drop results. Lower leaves die, sometimes leaving a tuft of wrinkled leaves at the top of the plant with some of the dead leaves clinging to the stem, resulting in a "palm tree" appearance. Vein banding and leafdrop streak are usually more severe in current-season infections. Plants

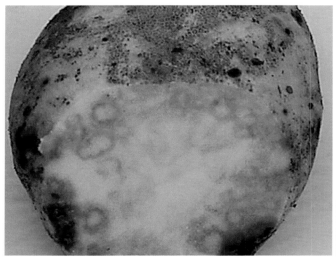

Necrotic strains of PVY cause the development of necrotic arcs or rings on the surface of tubers. The necrosis extends into the tuber flesh.

PHILIP B. HAMM

that grow from infected seed tubers may be dwarfed and may have mottled, wrinkled, and brittle leaves. They develop milder vein banding symptoms. Tuberborne infection by some strains causes severe browning (necrosis) of leaves and leaf drop. In Ranger Russet, tuberborne infections by some PVY strains kill plants, while other strains stunt the growth of plants.

Plants that are infected with PVY may develop a mosaic mottling that is similar to symptoms caused by PVA, but the infection usually appears as smaller, more numerous discolored areas. Mottling may be masked at temperatures below 50°F (10°C) or above 70°F (21°C). When PVY and PVX are present in the same plant, they may cause a severe wrinkling of the leaves known as rugose mosaic; veins are sunken and interveinal areas are raised. Milder rugose symptoms may be caused by strains of PVY or PVX alone.

Strains of PVY that cause necrosis of foliage and stems may also cause the development of necrotic arcs or rings on the surface of tubers or internal blemishes. External blemishes cause the skin to crack over the arcs and rings. Depending on cultivar and virus strain, the necrosis may or may not extend into the tuber flesh.

PVY has a wide host range that includes tomatoes, peppers, eggplants, legumes, nightshades, pigweeds, and many other weed species. The most important overwintering source of PVY is infected potato tubers. Some weed hosts may serve as overwintering sources, but their significance is not known. The virus is transmitted by at least 25 different aphids, the green peach aphid being the most efficient. However, large populations of other aphids that probe potatoes but are not hosted by potatoes could transmit as much virus as small numbers of green peach aphid. To minimize losses to PVY, plant certified seed tubers, eliminate cull piles,

PVY causes vein banding, the development of brown to black streaks along veins on the undersides of leaflets.

control potato volunteers and weeds, and prevent buildup of aphid populations. Because of the nature of the virus and the way it is transmitted, PVY is extremely difficult to control.

Corky Ringspot
Tobacco rattle virus (TRV)

Corky ringspot in potato tubers is caused by TRV, which is transmitted by stubby-root nematodes, *Paratrichodorus* spp. Corky ringspot symptoms are sometimes called "spraing" or "sprain." The disease occurs in certain areas of California, Colorado, Oregon, Idaho, and Washington, and is controlled by planting virus-free seed tubers, eliminating weed hosts of the virus, controlling stubby-root nematode, and avoiding fields with a history of corky ringspot.

Corky ringspot symptoms in tubers vary depending on virus strain, potato cultivar, temperature, and time of infection. Two types of symptoms are most common. The first is visible as rings of dark brown corky layers that alternate with rings of healthy tissue. These rings may originate at surface lesions caused by nematode feeding. Similar symptoms can also be caused by necrotic strains of PVY and *Alfalfa mosaic virus* (AMV). The second type of symptom is characterized by small, brown flecks that are diffused through the tuber. This second type of symptom may be confused with symptoms caused by certain strains of AMV, which are more common in some growing areas than corky ringspot, or a number of diseases or disorders characterized by browning in tubers, such as net necrosis caused by PLRV, internal brown spot, and Verticillium wilt. In White Rose and Kennebec, corky rings ¼ to ½ inch (6 to 12 mm) in diameter appear in mature tubers, while in Russet Burbank tubers, symptoms are spots,

arcs, or rings of brown tissue. The presence of TRV is difficult to confirm with standard serological techniques, but it can be confirmed using polymerase chain reaction (PCR).

Plants can become systemically infected with TRV. Aboveground symptoms are caused by certain strains of the virus and occur rarely, when plants grow from infected tubers or when sprouts are infected before they emerge. Some sprouts from a seed piece form stunted stems with leaves showing patterns of yellow lines, while other sprouts from the same seed piece grow normally.

Corky ringspot occurs on potatoes grown in coarse, sandy soils where the stubby-root nematode is present. The nematode acquires TRV by feeding on the roots of infected plants and spreads the virus to other plants as it feeds. Adult nematodes that acquire TRV remain infective for life, which can be as long as 2 years. A large number of weeds are hosts for TRV, including nightshades, pigweeds, shepherd's-purse, purslane, common cocklebur, birdsrape mustard, and sunflower. Many crop plants can be infected by the virus without showing disease symptoms. Young potato roots and tubers are infected with TRV when virus-infected nematodes feed on them. The virus is also transmitted in infected tubers and the seeds of hairy nightshade, purslane, and common cocklebur, any of which can introduce corky ringspot to a new area, where it will become established if the stubby-root nematode is present.

For control of corky ringspot, plant certified seed tubers, eliminate weed hosts, and control stubby-root nematodes (discussed in the seventh chapter, "Nematodes"). Crop rotations are ineffective for control of this disease because both the virus and nematode have broad host ranges. It is possible that a rotation crop of alfalfa will eliminate the virus from

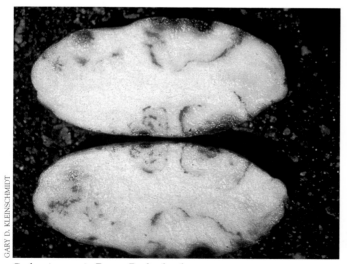

Corky ringspot in Russet Burbank tubers may appear as arcs or concentric rings of brown tissue.

Patterns of yellow lines may develop on the foliage of some sprouts growing from seed tubers infected with TRV.

AMV causes a pale or bright yellow mottling of leaves, called calico.

a stubby-root nematode population, but the alfalfa must be kept weed free. When growing processing potatoes or seed potatoes, either avoid fields with a history of corky ringspot or be certain to use effective preplant or postplant nematicide treatments. Many seed certification programs have a zero tolerance for TRV. In Idaho, seed potatoes cannot be grown for certification in a field that has a history of corky ringspot.

Calico
Alfalfa mosaic virus (AMV)

AMV causes the striking yellow symptoms on leaves known as calico that can be seen wherever potatoes are grown. Usually only an occasional plant is affected and the disease is unimportant, but it can become a problem when potatoes are planted next to alfalfa or clover, the normal hosts for this virus. Some strains of AMV cause cell death, or necrosis, rather than the pale to bright yellow mottling or blotching of leaf surfaces. Necrosis occurs mainly in the stems and can move into tubers, where it appears throughout as dry, corky areas or rusty brown patches. Necrosis may also occur in tubers of plants with characteristic calico symptoms. Tuber necrosis caused by AMV may be confused with corky ringspot, but plants that have tubers with corky ringspot symptoms rarely show any aboveground symptoms.

AMV is nonpersistent in at least 16 different aphid species, including the green peach aphid. Transmission of the virus is most likely to occur when aphids migrate into potato fields after an alfalfa field has been harvested. The pea aphid, *Acyrthosiphon pisum*, is probably the most important vector

because it is most likely to migrate from mowed alfalfa into potato fields. Aphids do not transmit the virus between potato plants, but the virus can be carried in infected tubers. To control AMV, plant certified seed tubers and avoid planting near alfalfa or clover, especially downwind. Older alfalfa fields are much more likely to be infectious because of virus buildup. When growing seed potatoes, remove calico plants and their tubers as soon as they appear.

Curly Top
Beet curly top virus (BCTV)

The beet leafhopper, *Circulifer tenellus*, transmits BCTV in a persistent manner. BCTV infects a wide range of other crop plants besides potatoes, including tomatoes, peppers, sugar beets, beans, squash, and melons. Both the virus and the leafhopper survive on a large number of wild plants. Curly top is considered of little importance in most potato-growing areas, but it has been common in southwestern Idaho, California's Kern County, New Mexico, and Utah.

Curly top symptoms include dwarfing, yellowing, and rolling of upper leaves. Leaves near the growing point develop yellow margins and become elongated, cupped, twisted, and rough. Outer leaflets are rounded at their tips, and their veins remain green while the rest of the leaflet turns pale yellow. Aerial tubers may form. The virus will infect tubers, and if they are planted they fail to emerge or develop stiff, erect stems that are pinched together at their tips and have leaflets that do not unfold completely.

No specific controls are used for this disease in potatoes. In areas where curly top is a problem in other crops, spray programs to control beet leafhoppers in native vegetation have reduced the incidence of curly top in potatoes and tomatoes.

Beet Leafhopper Transmitted Virescence, Aster Yellows, and Witches' Broom Phytoplasmas

The phytoplasmas that affect potatoes are transmitted by leafhopper vectors and infect a wide range of plants. Until recently no specific controls were necessary for phytoplasma diseases in potatoes, but special controls for leafhoppers may now be needed in some areas to prevent spread of phytoplasma diseases into potatoes, particularly in the Columbia Basin.

Symptoms and Damage

Phytoplasmas cause a yellowing or purpling and upward curling of potato foliage that resembles symptoms caused by PLRV. The discoloration caused by phytoplasmas tends to be more intense. Aerial tubers usually form in the axils of stems, and elongation of axial buds is common. The aster yellows phytoplasma may cause necrosis in tubers produced on plants with current-season infections. This necrosis may be confused with net necrosis caused by PLRV, but it tends to be scattered throughout affected tubers rather than being more concentrated in the vascular ring and at the stem end of the tuber, as is leafroll net necrosis. The disease known as "purple top" has been shown to be caused by both beet leafhopper transmitted virescence agent (BLTVA) and the aster yellows phytoplasma.

Infected tubers may fail to produce plants or produce plants that are severely stunted and fail to form tubers. Tubers infected with the witches' broom phytoplasma develop into plants with numerous stems and small, round leaves. These plants produce large numbers of pea-sized tubers. Because plants grown from tubers infected with phytoplasmas produce no useful tubers, phytoplasma diseases tend to be self-eliminating during seed production. However, the presence of phytoplasma disease symptoms in seed fields may prevent certification. Also, if a seed field is infected late in the season and tubers from that field are used for a commercial planting the following year, the commercial planting will have a higher incidence of disease.

Seasonal Development

Phytoplasmas are similar to bacteria. They are transmitted to potatoes from other hosts by several species of leafhoppers. Leafhoppers acquire phytoplasmas when feeding on other hosts and can infect other plants after a period of time, from a few hours to weeks. There is no significant secondary spread within potato fields because potatoes are not a preferred host for the leafhopper vectors, and they do not stay on potato plants long enough to acquire a phytoplasma. Phytoplasmas can be carried in some tubers produced on plants with late current-season infections.

The beet leafhopper, *Circulifer tenellus*, is the only confirmed vector of the BLTVA. Aster yellows is transmitted by the aster leafhopper, *Macrosteles quadralineatus* (= *M. fascifrons*). Witches' broom is transmitted by leafhoppers of the genus *Scleroracus*. Hosts for these leafhoppers and the phytoplasmas include a wide range of weed and crop species. Potatoes generally are not a preferred host for the leafhoppers, but the leafhoppers will migrate into potato fields when their preferred hosts are harvested or dry.

The beet leafhopper survives winter as adult females on winter annual weeds, primarily mustard family weeds and filaree. Mustards also are host for BLTVA. A new generation is produced in spring, and these migrate to summer hosts when mature and winter hosts die. In the Columbia Basin, this migration usually occurs in May and June. Preferred summer hosts include Russian thistle and kochia. If preferred hosts are not available, the adult leafhoppers will move into

Beet leafhopper transmitted virescence agent causes discoloration and upward curling of potato leaflets. Similar symptoms may be caused by other phytoplasmas and by PLRV.

potatoes and feed on potato plants long enough to transmit BLTVA, but do not reproduce on potato.

Management Guidelines

The incidence of phytoplasma diseases is kept to a minimum by using certified seed tubers and effectively controlling leafhopper vectors. Controls applied for aphids in seed fields and commercial fields usually keep leafhoppers controlled. Additional insecticide applications, especially early in the season, may be needed in some situations if the area has a history of phytoplasma problems, monitoring programs indicate that leafhopper flights are occurring, and transmission of phytoplasmas is a potential threat. Check with local authorities for the latest information on leafhopper control in your area. Information on leafhoppers and transmission of phytoplasma diseases in the Columbia Basin can be found at the Washington State University Web site Hotline for Aphids on Potatoes (www.wsu.edu/~potatoes).

Physiological Disorders

Environmental factors including improper cultural practices, high or low temperatures, and inadequate, excess, or uneven levels of soil moisture or nutrients may disrupt normal tuber growth and development and cause undesirable disorders. Disorders not caused by biological agents (microorganisms or insects) are called physiological or abiotic disorders.

Physiological disorders can affect the internal as well as external appearance of tubers. Tubers used for seed, fresh market, or processing may be affected; certain disorders are more important if tubers are intended for one of these markets than if they are intended for the others. Some cultivars are more susceptible than others to particular disorders. Losses from many physiological disorders can be avoided or minimized by using good cultural practices, particularly those that provide uniform and adequate soil moisture and fertility throughout tuber initiation and growth. However, the causes of some disorders are unavoidable or unknown. A number of publications on physiological disorders are listed under "Diseases and Disorders" in the suggested reading.

Secondary Growth

When environmental stresses have temporarily slowed or halted tuber growth, resumption of growth may cause non-uniform tuber development called secondary growth. Hail or frost injury to leaves, as well as high and low soil and air temperatures, low moisture, an imbalance in fertility, or combinations of these factors can interrupt normal tuber growth. Potato cultivars that produce long tubers tend to be more susceptible to secondary growth, but no cultivars are completely resistant.

Because all the unfavorable conditions that contribute to secondary growth disorders cannot be controlled, these disorders cannot be avoided entirely. However, attention to good cultural practices can minimize secondary growth. Plant for uniform stands to ensure desired tuber set: use correct spacing for the cultivar, take steps to eliminate skips and doubles, and follow proper handling procedures to avoid seed piece decay. Maintain levels of soil fertility and moisture that support uniform plant development throughout the growing period to maintain consistent tuber growth. Irrigation to maintain recommended soil moisture levels during hot weather is critical, but excessive soil moisture must be avoided. Avoid excess nitrogen, which causes excessive vine growth, especially later in the season.

Tuber Malformations. Several different types of tuber malformation can occur when growth resumes after being stopped. Each type has a descriptive name. Knobby tubers are formed when secondary growth occurs at one or more lateral buds. Growth in a longitudinal direction results in bottleneck, dumbbell, or pointed end; the type of symptom is determined by the time at which the growth interruption occurs. Growth interruption early in tuber development results in bottleneck or pointed stem ends, midseason interruption results in dumbbell, and late-season interruption tends to cause tubers with pointed bud ends. The constricted ends of pointed-end tubers often have abnormally high levels of reducing sugars and develop a symptom called translucent end or sugar end.

Knobby tubers are formed when secondary growth occurs at lateral buds.

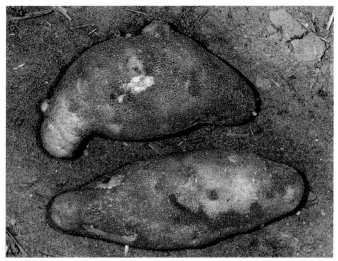

Pointed end, which is a constriction of either the stem or bud end of a tuber, results when secondary growth occurs early or late in the season.

Heat Sprout. Secondary growth may occur as sprouts that develop from buds on stolons or tubers. The secondary sprouts elongate and may remain underground or may emerge and develop into leafy stems. Heat sprout often occurs during periods of high soil temperatures and when tubers are close to the soil surface.

Tuber Chaining. Secondary growth may occur from a primary tuber as a series of tubers along a stolon. Tuber chaining frequently occurs after high soil temperature.

Little Tuber. Another form of secondary growth is little tuber, also called no top. This disorder involves the formation of many secondary daughter tubers from a planted seed tuber without foliage development. The daughter tubers may develop directly from buds or may be formed on short stolons that emerge from buds on seed tubers. The tubers are sometimes referred to as submarine tubers. Little tuber can be associated with improper storage handling of seed tubers. The disorder occurs most frequently when physiologically old seed tubers are planted, when seed tubers are planted into cold soil (below 50°F [10°C]) after they have been stored at a temperature of 68°F (20°C) or higher, or when seed tubers are placed in cold storage after sprouting and then planted at a later date. To control little tuber, store seed at 38° to 40°F (3.3° to 4.4°C), avoid using seed tubers that have been stored a long time, and do not plant older seed tubers in cold, dry soil.

Translucent End
Jelly End

The disorder translucent end, which is also called sugar end, glassy end, or incipient jelly end, occurs when the stem end of a tuber develops high reducing sugar levels. Usually less than 2 inches (5 cm) of the stem end is affected. The tuber tissue appears water-soaked or translucent. Tubers with translucent end produce french fries or chips that are undesirably dark in color because the heat during processing causes reducing sugars to react with amino acids, forming dark compounds. Translucent end tissue may become soft and jellylike, a symptom called jelly end. Jelly end may develop in the knobs or pointed ends of malformed tubers. The jelly end tissue may begin to decay, a condition called jelly end rot.

Translucent end and jelly end are associated with tubers showing secondary growth symptoms. Starch levels decrease in the stem ends of tubers undergoing second growth, and reducing sugar levels increase. Use the same cultural practices that reduce secondary growth to avoid translucent end and jelly end. Maintain adequate soil moisture, especially during tuber initiation and early tuber bulking. Maintain adequate fertility and avoid overfertilization, especially with nitrogen. Also avoid compacted soil conditions.

Growth Cracks

Sudden, rapid increases in the volume of internal tuber tissue cause growth cracks, or splits in the skin of tubers. The splits usually run lengthwise, and exposed tuber tissue heals, leaving fissures in the tuber surface. Tubers with external growth cracks are usually unacceptable for fresh market.

Irregular moisture conditions are often responsible for growth cracks. A heavy rain or irrigation following a dry period causes a rapid increase in growth activity and water uptake. The resulting internal pressure may split the skin. Fertilizer applications that cause rapid tuber growth may also cause growth cracks. When conditions favoring growth cracks occur, the incidence of growth cracks is greater where plant spacings are wide or if fewer tubers are set per hill.

To minimize problems with growth cracks, provide uniform growing conditions by maintaining adequate soil moisture and fertility throughout the growing season. Use correct seed piece spacing to obtain adequate stands. Cultivars vary in their susceptibility to growth cracks.

Surface Abrasions

Tubers with surface abrasions have a feathery, ragged appearance, symptoms referred to as skinning or feathering. Most abrasions usually occur at the bud end of the tuber. Under proper wound-healing conditions, surface abrasions will heal. However, if exposed to high temperature and dry air, tubers lose weight, a scald symptom may develop, and the injured areas may turn dark and become susceptible to rot. Tubers of red-, white-, and yellow-skinned cultivars are more susceptible to abrasions than those of russet cultivars.

Surface abrasions are caused by rough handling of immature tubers. If tubers are harvested before their skins have fully developed, they are more susceptible to abrasion. Allow tubers to mature properly by letting soil fertility and moisture decrease before vinekill and waiting 2 to 3 weeks between vinekill and harvest. Use careful harvesting and handling procedures, and always cover tubers during transit to prevent wind and sun damage.

Bruising

Three different types of bruising occur. Blackspot bruise and shatter bruise are caused by the bouncing of tubers against hard surfaces during harvesting and handling. Pressure bruise results from the pressure of other tubers or container surfaces during storage when tubers lose weight. Bruise damage is reduced by following careful harvesting and storage procedures.

Blackspot Bruise

Blackspot bruise appears as small to large dark discolored areas beneath the potato skin. Damage is usually not visible unless tubers are peeled. Tubers are initially bruised during harvesting and handling. Injured tissue first turns pink, then red, then darkens. When the temperature is about 70°F (about 21°C), discoloration begins about 6 hours after injury and is complete within 24 hours; discoloration occurs sooner at higher temperatures. Stem ends of tubers, dehydrated tubers, larger tubers, and tubers with higher specific gravity are more susceptible to blackspot bruise. Inadequate potassium levels increase the potential for blackspot bruise. Cultivars vary greatly in their susceptibility.

Shatter Bruise

An impact of sufficient strength to rupture the tuber skin and several layers of cells beneath it causes shatter bruise. The

Growth cracks are splits in the tuber surface caused by rapid expansion of tuber tissue. They heal over but make the tuber unacceptable for fresh market.

Blackspot bruises are small to large dark areas that develop just underneath the skin where a tuber has bumped against a hard surface.

resulting small splits in the skin may not be visible at first but are easy to see after the injured tuber tissue dries out. The margins of shatter bruises may show the same discoloration as blackspot bruises, but the injuries are visible on the tuber surface and usually penetrate deeper than blackspot bruises. Tubers are susceptible to shatter bruise when their temperature is low, 45° to 50°F (7.2 to 10.0°C); tubers with higher water content are more susceptible. "Thumbnail cracks" are shallow, curved cracks in the tuber surface that are caused by handling cold (< 45°F), turgid tubers. Thumbnail cracks are shallower than shatter bruises, but are caused by milder impacts. Even careful handling of tubers that are colder than 45°F (7.2°C) can result in thumbnail cracks.

Shatter bruises are small splits in the tuber skin. They are most visible after the injured tissue dries out.

Thumbnail cracks are shallow, curved cracks in the tuber surface that are caused by handling tubers that are turgid and colder than 45°F (7.2°C).

Controlling Blackspot and Shatter Bruise

Blackspot bruise and shatter bruise are minimized by careful harvest of tubers that are not cold and that have the correct water content. To provide the best tuber hydration at harvest, follow good cultural and pest control practices to ensure adequate vine and root development. Make sure potassium nutrition is adequate by using a preplant soil test and applying the recommended amounts of potassium for your area. Allow tubers to mature after vines are killed. Maintain adequate soil moisture until harvest; avoid excess irrigation, which can cause disease, but keep soil moisture above 60% of field capacity. Avoid harvesting when tuber pulp temperatures are below 45°F (7.2°C). When night temperatures are below 50°F (10.0°C), harvest during the warmest part of the day. Use harvesting and handling procedures designed to reduce bruising; consult the suggested reading for specific recommendations to reduce bruising during harvest operations. You can use a bruise test to identify where in a handling operation bruising is occurring. This information may help identify changes in operations that could decrease bruising.

Bruise Test. Blackspot and shatter bruises can be detected more quickly using heat or hot water to hasten the development of bruise symptoms. This may be useful for determining where bruising injuries are occurring or whether modifications in harvesting or handling procedures have reduced bruising.

Heat. Both shatter and blackspot bruise can be detected more quickly by heating tubers. Hold tubers at 90° to 95°F (32° to 35°C) for 12 hours, then inspect for bruise symptoms.

Hot Water. The hot water test detects both shatter and blackspot bruise. Place tubers in hot water (140°F [60°C]) for 10 minutes. Remove them, let them stand for 6 to 7 hours, then inspect them for bruise symptoms.

Pressure Bruise

Pressure bruises are flattened or sunken areas on the surfaces of tubers that result from low humidity in storage. When humidity is below 95%, pressure from storage surfaces or other tubers causes depressions in the tuber surface. Affected tissue loses water and the depressions persist if the humidity remains below 90%. Tuber quality is reduced. Some cultivars are more susceptible to pressure bruise; round white cultivars are more susceptible than Russet Burbank. To minimize pressure bruising, allow tubers to mature properly before harvest, follow good handling practices during harvest, provide good wound healing conditions at the beginning of storage—relative humidity of 95%, temperature at 50° to 55°F (10° to 13°C), good airflow—and keep the pile depth to 16 to 18 feet (4.9 to 5.5 m) or less. Use reduced pile depths when storing susceptible cultivars. Maintain proper humidity during storage by humidifying incoming air and making sure airflow is adequate but ventilation is not excessive.

Hollow Heart

The development of cavities in the pith area of tubers is known as hollow heart. The cavities may be very small or may take up the entire pith area, there may be one cavity or several, and cavities may develop near the stem end or near the bud end. In the case of stem end hollow heart, an area of dead brown cells develops before cavities form. This symptom is called brown center or incipient hollow heart and may develop while tubers are very small. Cells in the brown area later split apart, and cavities form in the middle of the brown areas. Bud end hollow heart usually appears late in the season and develops without any brown discoloration. The cavities are located nearer the bud end. A tan or brown layer of suberin that resembles tuber skin may form on the inside of hollow heart cavities.

Hollow heart is associated with periods of rapid tuber growth. Nutrient deficiency and stress that cause cell injury may contribute to the development of symptoms. Russet Burbank may develop brown center and stem end hollow heart or bud end hollow heart. Cool soil temperatures during tuber initiation induce brown center development; 5 days of 64°F (18°C) or lower day temperature and 50°F (10°C) or lower night temperature may induce brown center. Russet Burbank tubers initiated in wet soil may be more susceptible to hollow heart.

To decrease the incidence of hollow heart, use close plant spacings and large seed pieces that will ensure a good stand with an adequate number of stems per hill. This provides full tuber set and reduces the likelihood of periods of rapid tuber growth. Keep soil moisture and fertility adequate to maintain uniform tuber growth rates. When tubers are forming, irrigate carefully during cool weather to avoid wet soil and apply fertilizer carefully.

Internal Brown Spot

Internal brown spot consists of small, round or irregularly shaped, light tan or reddish brown corky spots that are scattered throughout the tuber tissue. This disorder is also known as internal browning, internal rust spot, physiological internal necrosis, chocolate spot, and internal brown fleck. Internal brown spot lesions may occur anywhere within the tuber, but they seldom occur in the cortex, outside of the vascular ring. The lesions may develop anytime during the season or in storage.

The cause of internal brown spot is unknown. However, this disorder is associated with certain conditions: hot, dry weather, low soil moisture, sandy soils (compared to soils

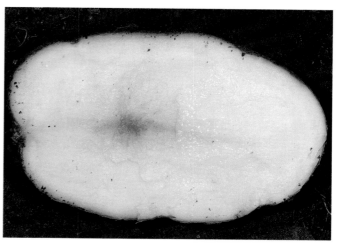

Brown center, or incipient hollow heart, is an area of dead, brown cells that develops in the center of tubers.

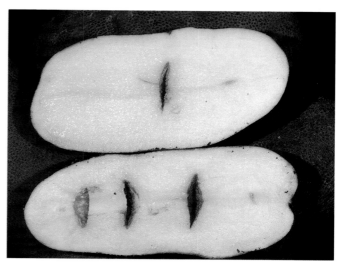

Hollow heart is a cavity that forms in the center of a tuber. It may or may not be surrounded by an area of dead, brown cells.

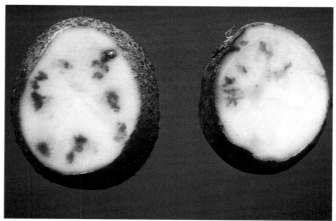

Internal brown spot consists of small spots of light tan or reddish brown corky tissue that are scattered throughout the tuber tissue.

ARTHUR KELMAN

high in organic matter), uneven growth rates due to temperature or moisture conditions, and low calcium levels in tuber tissue. When calcium fertility or availability is low and soil temperatures are high, the incidence of internal brown spot increases. Some cultivars are more susceptible to internal brown spot, and the disorder tends to occur more frequently in larger tubers.

To reduce the occurrence of internal brown spot, provide proper soil fertility and moisture and maintain uniform growing conditions throughout the growing season. Allow tubers to mature between vinekill and harvest, but do not allow tubers to lie in hot, dry soil. Cut samples of tubers at harvest to evaluate the incidence of internal brown spot.

Infection by certain viruses can cause necrotic symptoms inside tubers that may be confused with those that have a physiological origin. Tubers with necrotic symptoms should be tested for the presence of virus to be sure of the cause.

Internal Heat Necrosis

Symptoms of internal heat necrosis are similar to those of internal brown spot but occur mainly in the area of the vascular ring. Internal heat necrosis is caused by high soil temperatures between the time vines begin to die and harvest time. Tubers in the upper 2 inches (5 cm) of soil are most likely to develop symptoms, which appear at or near harvest. Use irrigations, mulching, or increased soil cover to reduce exposure of tubers to high temperatures. If you use irrigation during hot weather to cool the soil, be sure to keep the soil moisture below field capacity. Wet soil during hot weather can result in black heart or tuber rot. If possible, schedule harvest to avoid high soil temperatures.

Vascular Discoloration

Killing vines when soil moisture is low or vine-killing frosts that occur in dry weather may cause vascular discoloration in tubers. The discoloration occurs in the region of the vascular ring and may be light tan, reddish brown, or dark brown. Vascular discoloration may be confined to an area near where the stolon attaches to the tuber, or it may extend throughout the tuber. This disorder is also known as stem end discoloration, browning, or necrosis, and vascular browning. Vascular discoloration may be present at harvest or may develop during the first 2 months of storage.

Physiological vascular discoloration can be difficult to distinguish from similar symptoms caused by PLRV or *Verticillium*. PLRV causes net necrosis in the tubers of some cultivars, a discoloration that is restricted to phloem tissue but extends throughout affected tubers. Tubers with leafroll net necrosis can be distinguished by laboratory detection of the virus or by planting the tubers, which grow into plants with chronic leafroll symptoms. Tubers on plants infected by *Verticillium* may develop discoloration of xylem tissue; however, it is confined to the stem end and is usually darker than physiological vascular discoloration. Vines usually show symptoms of Verticillium wilt before they are killed, distinguishing the biotic disease from the physiological disorder.

It is not known how to control this physiological disorder. It has not been possible to consistently cause stem end discoloration under controlled conditions. However, precautions that should be taken include maintaining adequate soil moisture during vine-killing operations and avoiding killing stressed vines.

Black Heart

Black heart is a dark gray, purple, or black discoloration of the tuber pith area that is caused by insufficient oxygen. Brown, purple, or black patches may be visible on the tuber surface on rare occasions. Cavities resembling hollow heart occasionally develop, but unlike hollow heart they are surrounded by gray or black layers. These cavities are sometimes referred to as "cat's eye." Black heart may develop in the field or in storage where reduced gas exchange occurs.

Black heart results when internal tuber tissue does not receive enough oxygen to support respiration. In the field, black heart is usually caused by waterlogged soil or extremely high soil temperatures. Waterlogging prevents oxygen from getting to the tubers; high soil temperatures increase respiration rates so that oxygen cannot reach internal tuber tissues fast enough. In storage or transit, black heart results from poor ventilation, improper heating of tubers during shipment to prevent freezing, or prolonged storage at temperatures near freezing.

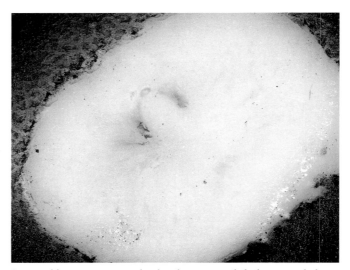

Stem-end browning or vascular discoloration is a light brown to dark brown discoloration of the vascular tissue at the stem end of tubers. It can be caused by the Verticillium wilt fungus or physiological stress during vinekill.

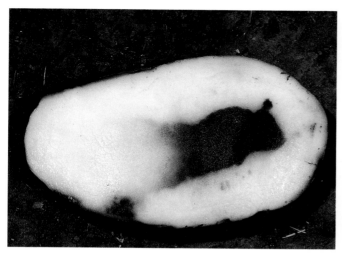

Black heart is a dark gray, purple, or black discoloration in the center of a tuber exposed to waterlogged soil or high soil temperatures.

If tuber tissue is frozen, a dark line develops between injured and uninjured tissue. Affected tissue becomes water-soaked and is easily seen on the surface. Frozen tissue usually disintegrates.

Potato cultivars differ in their susceptibility to black heart, but precautions must be taken with all cultivars. Make sure that soil drainage is good and that there are no spots in the field where water collects. Do not leave tubers in or on hot soil. Limit the amount of dirt and debris in storage, and store tubers at the proper temperature with proper airflow and air distribution. Avoid exposing tubers to temperatures above 90°F (32°C).

Low-Temperature Injury

Temperatures below freezing may injure tubers. The severity of the damage depends on the temperature and duration of exposure. The flesh of tubers exposed to temperatures slightly below freezing may turn gray or reddish brown, a symptom called mahogany browning. Tubers exposed to temperatures below 28°F (−2.2°C) for extended periods sustain serious injury. If cells are not frozen, symptoms will not be visible on the tuber surface, but internal cells die and tissues turn brown, causing the symptoms of chilling injury. If ice crystals form in the cells, freezing injury results. Freezing injury in the field usually occurs on the end or side of the tuber that is closest to the soil surface, usually the bud end, and symptoms are easily seen on the tuber surface. Tissues at the stem end and in the vascular ring are more sensitive to low-temperature injury. Affected tissues become liquid, and the vascular ring may break down completely. Frozen tissue may dry out and become tough or chalky. Use proper hilling to help protect tubers from freezing temperatures. Whenever possible, harvest and store tubers before severe frost is likely. Use high airflow to reduce storage losses in lots that were exposed to freezing temperatures in the field. Exposure of tubers to freezing temperature during transit or in storage results in tissue damage to exposed surfaces.

Greening

Tubers that are exposed to sunlight in the field, or enhanced light in storage or at market, may develop greening. The light causes the green pigment chlorophyll to form just beneath the skin, giving tubers a green color. This symptom may also be called greened tubers, sun green, or virescence. The chlorophyll is harmless, but levels of other compounds, called glycoalkaloids, also increase in the green tissue. They are mildly toxic and give tubers a bitter taste, making them unpalatable. Tubers with greening are unacceptable for fresh market or processing. White-skinned cultivars such as Cal White, White Rose, and Yukon Gold are more susceptible to greening. Use careful hilling operations early in the season to make sure tubers remain covered with soil throughout the season. Certain cultivars form tubers shallowly and require deeper planting or higher hills. Broad hills are better than narrow, peaked hills. Avoid fluorescent lights in storage. The rate of greening is slower at lower storage temperatures.

Enlarged Lenticels

Excessive moisture in contact with tuber surfaces can cause enlarged lenticels (the small pores present in the skin of tubers that provide air exchange), also known as water spots or water scab. Tissue beneath lenticels swells, bursts through the protective suberin layer beneath the lenticels, and forms raised masses of corky tissue on the tuber surface. Enlarged lenticels resemble scab lesions, but are smaller and lighter in color; they are susceptible to infection by decay organisms. Enlarged lenticels are usually caused by excessive soil moisture in the field, but they may also form in storage if free moisture is allowed to remain on tuber surfaces. Avoid overwatering or ponding of water in the field, and use forced-air ventilation to dry wet tubers as quickly as possible after harvest. Avoid condensation on tubers during storage by adequate storage insulation and by using proper ventilation, humidity, and temperature controls. Avoid tuber pulp temperatures of 65°F (18.3°C) or higher.

Elephant Hide

The development of thick, coarse russeting, primarily on Russet Burbank tubers, is called elephant hide or alligator hide. It is occasionally seen on other russet cultivars. Furrowing and cracking of a thickened periderm gives affected tubers the appearance of an elephant's hide. A similar condition occurs on smooth-skinned cultivars, where it is called fishy skin or turtle back. The causes and control of this disorder are not known. It may be the result of unfavorable environmental factors, contact with decaying organic matter or with salts in the soil, or certain soil-applied chemicals. The sprout inhibitor maleic hydrazide has been associated with the occurrence of elephant hide type symptoms in rare cases.

Aerial Tubers

Severe restriction of carbohydrate movement from vines into growing tubers causes the formation of small tubers at the base of stems or in leaf axils, where leaves join stems. These tubers are called aerial tubers, stem tubers, or air tubers. One or several tubers may form on a plant, often close to the ground. Aerial tubers are green to purple in color, usually small, and oddly shaped. Aerial tubers may be caused by vari-

ous disease pathogens, including *Pectobacterium*, *Rhizoctonia*, *Beet curly top virus*, and phytoplasmas; by insect injury to stems or psyllid toxin; or by mechanical injury or waterlogging of the soil.

Ozone Injury

The air pollutant ozone frequently reaches levels in southern California potato-growing areas that damage sensitive cultivars. Ozone injury also occurs in the southern San Joaquin Valley. Symptoms of ozone injury, frequently called "speckle-leaf," appear as tiny spots of dying tissue on upper leaf surfaces and bronzing on the lower surfaces. Yields may be reduced. If injury is severe, plants are defoliated. Avoid planting sensitive cultivars such as Centennial Russet and Red LaSoda in areas where ozone frequently reaches damaging levels. Damage is accentuated by low nitrogen levels in the soil.

Herbicide Injury

Potatoes can be damaged by herbicides applied during the season, by herbicide residues that carry over in the soil from previous crops or in seed tubers, or by drift from a neighboring area. Herbicide injury is characterized by yellowing foliage or distorted growth; browning or necrosis often, but not always, occurs. Laboratory analysis of soil, water, or plant residues may be required to verify injury related to herbicides. However, presence of herbicide residue is not necessarily proof that it caused the injury. Also check for the presence of diseases, especially viruses, that may cause similar symptoms. Analysis is easier if you can tell the labo-

Enlarged lenticels are raised masses of corky tissue that protrude through the lenticels of tubers exposed to excessive moisture.

Elephant hide, or alligator hide, is the development of thick furrowed and cracked skin that occurs primarily on Russet Burbank tubers.

Aerial tubers may form at the base of stems or in leaf axils when disease or physiological stress interrupts the flow of carbohydrates from leaves to tubers. Pathogens that may cause aerial tuber formation include *Pectobacterium, Rhizoctonia*, BCTV, and phytoplasmas.

ratory what materials have been applied; keep records for all fields of all materials and rates used during rotations as well as on the potato crop.

The classes of herbicides that may damage potatoes are listed below, along with descriptions of the symptoms they cause. Use these descriptions to help identify herbicide injuries, but also check with local experts. More information and illustrations can be found in *Herbicide Drift and Carryover Injury in Potatoes* listed under "Diseases and Disorders" in the suggested reading.

Acetanilides: metolachlor

Metolachlor is soil-applied to potatoes before emergence and may cause injury under some circumstances. Injured leaves appear bronze; leaf margins curl up or down and turn brown. Leaf surfaces appear crinkled or roughened. Occasionally leaves are small and odd shaped with roughened surfaces.

Bromoxynil

Injury results when bromoxynil spray drifts onto potato foliage during the treatment of adjacent grain fields. Localized areas on sprayed leaves turn bronze, then yellow. Leaf margins may curl up and turn brown, and interveinal tissue may turn brown.

Dinitroanilines: pendimethalin, trifluralin

Dinitroanilines are applied to the soil before potatoes emerge. Injury may occur if sprouts grow slowly through treated soil that is cold and wet. Symptoms are more frequently caused by trifluralin than by pendimethalin. Developing shoots become thick and brittle, root formation may be slowed or inhibited, and leaves may become crinkled and cupped.

Ozone causes spots or patches of dead tissue on the leaves of sensitive potato cultivars. This symptom is often called speckle leaf.

Glyphosate

Drift of glyphosate spray onto potato foliage causes injury. Youngest tissue is affected first. Petioles bend downward and leaf margins curl slightly, giving growing points a wilted appearance. The youngest leaflets turn yellow. In severe cases older foliage is affected and plants may die. Tuber malformations sometimes occur, including growth cracks and second growth symptoms. A fold or crease may develop, usually at the bud end of tubers. Roughened surface indentations resembling grub damage may also develop. Seed tubers produced on plants exposed to glyphosate spray drift may fail to emerge when planted. Affected tubers produce numerous spaghetti-like sprouts from each eye.

STEVEN R. JAMES

Seed tubers from plants that received glyphosate spray drift produce numerous spaghetti-like sprouts from each eye. These sprouts usually fail to emerge.

ALVIN R. MOSLEY

The youngest leaves of plants injured by glyphosate appear wilted and turn yellow.

Tubers of plants impacted by glyphosate spray drift early in the season may develop folds or clefts and a rough surface.

Growth Regulators: 2,4-D, 3,6-DPA, dicamba, picloram, clopyralid, fluroxypyr

Fluroxypyr, dicamba, and 2,4-D injuries are caused by spray drift; 3,6-DPA, clopyralid, and picloram injuries result from drift or residue carryover in the soil. Dicamba residues are not a problem when the herbicide is applied at normal rates at least 6 months before potatoes are planted. Symptoms appear on actively growing tissue. Margins of the youngest leaves curl or the leaves fold up. Leaflets appear roughened or crinkled, and petioles bend downward. With severe injury, young stems are curled and leaves are thin and severely distorted or absent; these symptoms are often called "fiddleneck." Dicamba and picloram may cause tuber malformations. If dicamba spray injury occurs during tuber growth, tubers may be rounder than normal, a cleft may develop at the bud end, and the surface of the bud end may develop a roughness similar to elephant hide. Occasionally, bull's-eyes or circles develop around tuber eyes.

Imidazolinones: imazapyr

Sulfonyl ureas: chlorsulfuron, diuron, linuron, metsulfuron, tribenuron-methyl

Injury symptoms caused by diuron and linuron are similar to those of triazine injury. They are usually caused by residue carryover from previous crops, but may also result from spray drift. Symptoms caused by sulfonylureas and imidazolinones include stunting, yellowing and wilting of plants, stubby roots, tuber malformations, growth cracks in tubers, and sometimes elephant hide. Chlorsulfuron is generally not recommended for use in rotations where potatoes will be planted, because the waiting period for potatoes is at least 3 or 4 years. Very small quantities of sulfonyl ureas injure potatoes. Avoid using a sprayer that has contained sulfonyl urea for applying materials to potato crops.

Injury by plant growth regulator herbicides, such as 2,4-D, dicamba, or picloram, appears as roughened, crinkled leaves. With severe injury, "fiddleneck" develops—young stems are curled and leaves are severely distorted or absent.

Tubers of plants injured by dicamba or picloram may be rounder than normal with clefts at the bud end and rough surfaces.

Triazine herbicides, such as atrazine, hexazinone, simazine, or metribuzin, may cause yellowing of tissue along the leaf veins of some cultivars. In other cultivars leaf tissue turns yellow between veins.

Thiocarbamates: EPTC

Potato foliage may be injured if EPTC is applied at too high a rate or if applications overlap during preplant soil incorporation. Shoots appear twisted, and leaves become cupped, curled, or bent over.

Triazines: atrazine, hexazinone, metribuzin, simazine

Atrazine, hexazinone, and simazine usually cause damage when residue remains from treatment of a previous crop. Metribuzin may cause injury when high rates are used on potatoes, when it is applied to sensitive cultivars, or when poor growing conditions (cloudy, cool, wet weather) occur before or after it is applied. Metribuzin injury is more likely following foliar applications. Triazine injury symptoms from soil uptake appear in larger, older leaves; symptoms from a foliar application are confined to the leaves sprayed. Leaves turn yellow either along the veins or between the veins, depending on cultivar. If injury is minor, the yellow tissue turns green; if severe, yellow tissue eventually dies and leaf margins take on a burned appearance.

Uracils: terbacil

Terbacil injury is usually caused by residue carryover from previous crops. Symptoms are similar to those caused by triazines.

Nutrient Deficiencies

Nutrient deficiencies may be caused by low levels of nutrients in the soil and by factors that reduce the ability of the potato plant to absorb nutrients. Excessive rain or irrigation can leach mobile nutrients such as nitrogen from the root zone, soil pH or nutrient imbalance can decrease the availability of certain nutrients, and disease may reduce the ability of roots to absorb nutrients. Deficiency symptoms include abnormal or reduced growth, discoloration of foliage, reduced yields, and improper tuber development. Plants deficient in nutrients are more susceptible to diseases and disorders such as Verticillium wilt, early blight, bruising, and internal browning. Use soil and tissue analysis to assess the nutrient status of your soil and crop, and follow fertilizer recommendations for your area and the cultivars you grow to avoid nutrient deficiencies. Identification of nutrient deficiencies may be difficult. Use the results of soil and tissue analyses and seek the advice of experts before taking corrective action.

Magnesium

Symptoms of magnesium deficiency develop on older, lower leaves. The interveinal tissue of leaflet margins and tips turns pale, and the discoloration progresses toward the center of the leaflet, where symptom expression becomes most severe. Discolored tissue turns brown and leaves become thick and brittle and roll upward. Magnesium deficiency usually occurs

ALBERT ULRICH

Magnesium deficiency causes yellowing of interveinal leaf tissue. Brown spots develop and leaves thicken and roll upward at their edges.

on acidic, sandy soils, from which the nutrient is easily leached; however, magnesium deficiency may also occur on finer-textured soils. Symptoms often appear after heavy rains, which leach magnesium from the root zone. High levels of potassium in the soil aggravate magnesium deficiency. Foliar sprays of magnesium sulfate or chelated micronutrients may correct deficiencies that appear during the season.

Manganese

Manganese deficiency symptoms develop on the upper part of the potato plant. Interveinal leaf tissue first loses its luster and turns pale. Discoloration progresses until the tissue becomes yellow or white. If manganese deficiency is severe, brown spots may develop along the veins of younger leaves. Manganese deficiency occurs most often in muck soils or calcareous, alkaline soils. Foliar applications of manganese sulfate or chelated micronutrients may correct deficiencies during the season.

Nitrogen

Potato plants deficient in nitrogen grow more slowly, form small, pale leaves, and turn yellow. Leaf veins stay green longer as leaves turn yellow. Older leaves are affected more severely. Nitrogen deficiency reduces yields and makes plants more susceptible to Verticillium wilt, early blight, and other diseases. Nitrogen deficiencies can be corrected by midseason fertilizer applications, but some loss of yield and tuber quality may still occur. Application of nitrogen to deficient plants stimulates regrowth and increases the chances of certain physiological disorders.

Leaves of plants with manganese deficiency turn pale. Brown spots devlop along the veins of younger leaves when the deficiency is severe.

ALBERT ULRICH

Phosphorus deficiency causes leaves to curl and turn a purplish color.

ALBERT ULRICH

Phosphorus

Early-season phosphorus deficiency causes leaves to curl and turn a purplish color. Phosphorus deficiency late in the season may cause premature maturation, increase disease susceptibility, and reduce yields. Phosphorus deficiencies are difficult to correct during the season. Use preplant soil analysis and results of petiole analysis from the previous potato season as guides to determine adequate phosphorus levels before and during planting. High-solubility phosphorus fertilizers can be applied during the season.

Potassium

Leaves of plants deficient in potassium develop a dark green to bluish color early in the season. Older leaves turn brown at their outer edges, become bronzed in appearance, and die early. Margins of leaves in the upper plant curl upward, and the leaflets become bronzed. Potassium-deficient plants are more susceptible to early blight and Verticillium wilt; their tubers are more susceptible to blackspot bruise and early blight infections. In-season applications can correct potassium deficiencies.

Sulfur

Yellowing of potato foliage throughout the plant may indicate a sulfur deficiency. The intensity of the yellowing varies from slight to severe. Leaflets may develop some upward curling. Sulfur deficiency may occur where sulfur levels are low in soil or in irrigation water.

Zinc

Potato plants suffering from zinc deficiency are stunted and have younger leaves that turn yellow and roll upward,

ALBERT ULRICH

Nitrogen deficiency causes leaves to turn a pale yellow green. Older leaves are usually affected more severely.

Leaves of plants deficient in potassium develop a dark green to bluish color. Upper leaves curl; lower leaves turn yellow at the edges, and brown patches develop.

symptoms that are similar to current-season leafroll and injury from some growth regulator herbicides. Bronze areas that later turn brown and die may appear on leaves. This symptom develops first on leaves in the middle of the plant and spreads throughout the plant later. Brownish spots may appear on stems and petioles. Zinc deficiency is most commonly seen on newly developed land and on alkaline soils. Excessive liming and high phosphorus levels accentuate zinc deficiency. Foliar applications of zinc chloride, zinc sulfate, or chelated micronutrients can correct zinc deficiency.

Young leaves of plants suffering from zinc deficiency turn pale and roll upward. Lower leaves develop bronze areas, turn brown, and die.

Nematodes

Nematodes are tiny roundworms that live in soil, water, and plant tissues. Many damage plants, but others feed on fungi, bacteria, or other nematodes. In the western states root knot nematodes cause the most damage to potatoes. Some root lesion nematode species, which are found in many potato fields, may increase the severity of early dying and may reduce yields if present in high numbers. Stubby-root nematodes are widespread in the western states; they transmit *Tobacco rattle virus* (TRV), which causes corky ringspot. The potato rot nematode has been reported in the West but occurs rarely. Potato cyst nematode does not occur in the western states, but growers should be aware of its symptoms because it might be introduced in the future.

Cropping history, soil sampling results, and past experience in local soils must all be considered when making nematode management decisions. Where soil populations of a nematode are high enough to cause unacceptable damage, plant nonhost crops or use fumigation or other soil treatments to reduce the nematode to levels that will not cause economic damage. Sanitation and certified seed help prevent spread to uninfested fields. Crop rotation with good weed control helps prevent established populations from increasing.

Root Knot Nematode
Meloidogyne spp.

Root knot nematodes occur throughout the potato-growing areas of the western states and are presently serious pests in California, Oregon, Washington, and Idaho. Most crop plants are susceptible to one or more species of root knot nematode. Several species of *Meloidogyne* are known to attack potato, and three of them are of major importance. All have similar life cycles, but the species differ in environmental requirements, host range, and some of the symptoms they cause. The distribution of each nematode species (Fig. 36) is determined primarily by its temperature requirements (Table 15). The Columbia root knot nematode, M. *chitwoodi,* and the northern root knot nematode, M. *hapla,* occur in the cooler growing areas; M. *chitwoodi* is more active than M. *hapla* at lower temperatures. The southern root knot nematode, M. *incognita,* is the predominant species in the San Joaquin Valley and southern California, Arizona, and New Mexico. *Meloidogyne hapla* is known to occur in these warmer areas, and occasionally fields will be infested with Javanese root knot nematode, M. *javanica,* or peanut root knot nematode, M. *arenaria.* Two or more species are often found together, but one usually predominates.

Table 15. Temperature Requirements of the Three Major Root Knot Nematode Species that Infect Potatoes.

	Soil temperature at root or tuber depth for species		
	M. chitwoodi	*M. hapla*	*M. incognita*
minimum for activity	43°F (6°C)	50°F (10°C)	65°F (18°C)
minimum for infection	43°F (6°C)	54°F (12°C)	65°F (18°C)
optimum for activity	68° to 77°F (20° to 25°C)	77° to 86°F (25° to 30°C)	77° to 90°F (25° to 32°C)

Losses from damage to tubers by some root knot species in some areas may be reduced by early harvest. In winter and spring production areas, damage may be avoided by harvesting before nematode populations build up to damaging levels. Losses due to Columbia root knot nematode may also be reduced by avoiding storage of tubers that may have been lightly infected. Root knot nematode populations may be managed with rotations to nonhost crops, green manure cover crops, dry fallowing, tillage to increase exposure to extreme temperatures, and nematicides.

Description and Biology

Root knot nematodes are endoparasites that must feed inside roots or tubers to reproduce. All common species attack a wide variety of crops and weeds (Table 16). Feeding by most species causes host plants to produce swellings, called galls or knots, around the feeding sites (Fig. 37). These abnormalities prevent normal uptake of water and nutrients.

Mature females, found only in roots or tubers, are pear shaped and up to about 1/16 inch (1.5 mm) long. They can sometimes be seen as tiny white "pearls" when infected roots are cut open or when slices of infected tubers are held up to the light. Males are worm shaped and under normal conditions are not abundant; they are not necessary for reproduction. Females lay eggs in jellylike masses on or just below the surface of infected roots and inside infected tubers. Tuber cells surrounding egg masses turn brown, and some of the

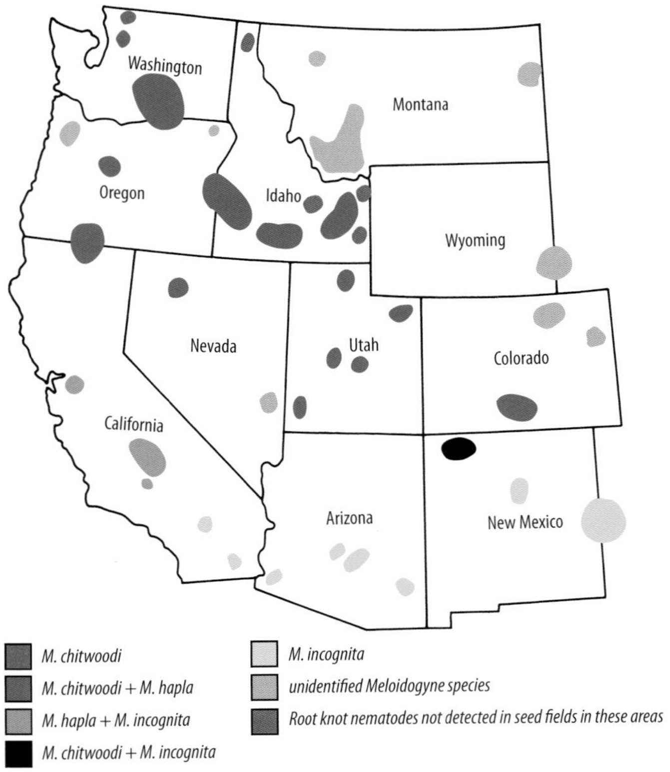

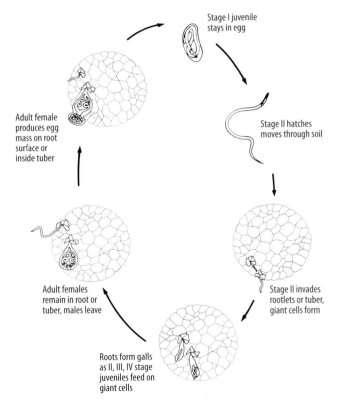

Figure 36. Distribution of major root knot nematode species in the potato-growing areas of the western United States.

Legend:
- M. chitwoodi
- M. chitwoodi + M. hapla
- M. hapla + M. incognita
- M. chitwoodi + M. incognita
- M. incognita
- unidentified Meloidogyne species
- Root knot nematodes not detected in seed fields in these areas

Figure 37. The root knot nematode spends much of its life cycle in roots or tubers. Second-stage juveniles invade new sites, usually near root tips or tuber lenticels or eyes, causing some cells to grow into giant cells where the nematodes feed. As feeding continues, tubers produce lumps; roots produce small galls in response to some species.

Life cycle labels:
- Stage I juvenile stays in egg
- Stage II hatches moves through soil
- Stage II invades rootlets or tuber, giant cells form
- Roots form galls as II, III, IV stage juveniles feed on giant cells
- Adult females remain in root or tuber, males leave
- Adult female produces egg mass on root surface or inside tuber

egg mass may also be brown. Eggs inside tubers can survive winter conditions; a small proportion of M. chitwoodi eggs can survive for more than 2 years inside tubers held at 34°F (1.1°C).

The worm-shaped juveniles that hatch from eggs can move as far as 6 feet (1.8 m) upward through moist soil, but they usually travel only short distances horizontally to find a host plant. They penetrate roots just behind root tips, where the root surface has not been suberized. The site of tuber infection is not known for certain; juveniles probably enter through lenticels or eyes, but they may also penetrate directly through the skin. Young tubers are not susceptible to infection by Columbia root knot nematode until lenticels have developed. Mature tubers with well-developed skin appear to be infected less than immature tubers. However, Columbia root knot nematode will continue to infect tubers as long as tubers remain in the ground. Inside the root or tuber, the nematode's salivary secretions contain enzymes and plant hormones that stimulate the formation of "giant cells," greatly enlarged cells that supply the nematode with food. Other cells may multiply to form swellings or galls. As they mature, females swell, become pear shaped, and lose the ability to move.

When susceptible host plants are available, a root knot nematode population increases at a rate that depends largely on soil temperature and moisture and the size of the initial population present in the spring. Populations usually decline by over 50% during the winter, but during mild winters populations may decline very little. Meloidogyne chitwoodi is active at lower temperatures than the other species, M. hapla and M. incognita; therefore, it builds up to damaging levels earlier in the season. In areas where M. incognita is the important species, most potatoes are planted in winter and harvested in late spring. Because this species does not infect until soil temperatures reach 65°F (18°C), tubers are usually harvested before the nematode can cause damage.

Root knot nematodes usually are concentrated in the upper 2 feet (0.6 m) of soil where feeder roots are abundant, but they can be found as deep in soil as roots penetrate; M. chitwoodi has been found as deep as 6 feet (1.8 m).

Meloidogyne chitwoodi can mature and lay eggs in stored tubers at temperatures as low as 43°F (6°C), thereby increasing damage during storage. An increase of tuber damage by other species in storage has not been reported.

Symptoms and Damage

Root knot symptoms occur on tubers and roots, but symptoms on tubers are more common and easier to see. Meloidogyne chitwoodi and M. incognita cause characteristic bumps or warts on the surfaces of infected tubers. Meloidogyne hapla causes swellings that are less distinct; it may be hard to distinguish these tubers from healthy ones just by observing the surface. Root knot nematodes are usually located in the tuber

Table 16. Selected Hosts of the Major Species of Root Knot Nematodes that Infest Potatoes.[1]

| | Root knot nematode species | | |
	Columbia M. chitwoodi	Northern M. hapla	Southern M. incognita
Crops			
alfalfa	+[2]	+	+
apple	-	+	+
barley	+	-	+
beans	+[3]	+	+
beet, table	-	+	+
carrot	+[2]	+	+
cole crops	+	+	+
corn	+[4]	-	+
cotton	-	-	+
cowpea	-	+	+
eggplant	-	+	+
grape	-	+	+
hops	-	-	0
lettuce	0	+	+
mint	-	+	-
melons	-	+	+
oat	+	-	+
onion	-[5]	+	+
peas	+	+	+
peppers	-	+	+
radish	-	+	+
strawberry	-	+	+
stone fruits	-	-	+
sudangrass	-[6]	-	-
sugar beet	-	+	+
tomato	+	+	+[7]
turnip	-	+	+
watermelon	-	-	+
wheat	+	-	+
Weeds			
barnyardgrass	-	-	-
bindweed	0	+	+
canada thistle	+	+	+
foxtails	-[8]	-	+
kochia	0	+	0
lambsquarters	-	+	+
mallow	0	+	+
mustards	-	+	+
nightshades	+	+	+
nutsedge	0	0	+
pigweeds	-	-	+
quackgrass	-	0	0
russian thistle	-	0	0
sowthistle	+	+	+

1. + = good host, - = poor or nonhost, 0 = unknown. Host ability may vary with cultivar; check with local authorities before selecting a rotation crop.
2. Alfalfa is a host for race 2 of M. chitwoodi, carrots are a host for race 1.
3. Lima bean is a nonhost for M. chitwoodi.
4. Popcorn and super-sweet cultivars are poor hosts.
5. Onion appears to be a host for M. chitwoodi in some fields.
6. Certain cultivars are good hosts.
7. Some cultivars are nonhosts for this species.
8. Green foxtail is a moderate host; yellow and meadow foxtails are poor or nonhosts.

Root knot nematodes cause various types of swellings, bumps, and warts on tuber surfaces, depending on the root knot species (tuber at left is uninfected; tuber at right is infected with Columbia root knot nematode). All species cause brown spots in the cortex of the tuber (below).

cortex, between the skin and vascular ring, but sometimes they penetrate deep into the tuber.

Brown spots develop in tubers around the egg masses of all root knot nematode species, which occur mostly in the outer ¼ inch (6 mm) and are best seen by carefully peeling off thin layers of tuber. The spots darken when tuber slices are cooked to make chips; therefore, tuber lots that are significantly infected with any root knot nematode species are unacceptable for processing. Warty tubers cannot be used for fresh market.

Root knot galls on feeder roots are usually small and difficult to see. *Meloidogyne chitwoodi* produces egg masses that appear as tiny round bumps on feeder roots of heavily infected plants. *Meloidogyne hapla* causes small, distinct galls and a proliferation of lateral roots around these galls. *Meloidogyne incognita* can cause more pronounced root galls, but it usually is not active in cool weather, when most potatoes are grown in the areas where it occurs.

Root knot nematode infections rarely cause aboveground symptoms in potatoes. However, plants severely infected by M. *incognita* may be stunted, turn yellow, wilt, or die if stressed for water.

Management Guidelines

Management decisions for root knot nematodes must be made well in advance of planting. Fumigation, when effective, works best in the fall, so it is important to keep records from the previous potato crop and to monitor in the season before planting. Check for nematode symptoms before harvest and keep records of where they are found; the field location of tuber symptoms in the last potato crop gives a good indication of where to sample for nematodes in the future. Egg masses on the surface of root samples taken at midseason

are easily seen by staining with Phloxine B. Some weed hosts can be good indicators of root knot nematode. *Meloidogyne chitwoodi* produces obvious galling on bull thistle and spiny sowthistle. *Meloidogyne hapla* produces distinct galls on dandelion; M. *incognita*, on nightshades and groundcherries. Symptoms on weeds or previous crops are a good sign that potatoes planted in the same soil are likely to be infected; however, the absence of root knot symptoms does not necessarily mean that root knot nematodes are not present or that a potato crop will not be damaged. Some crops and weeds either are not susceptible to infection or do not develop distinctive symptoms. Root knot nematodes may not be detectable in soil when no suitable host is present.

Tomatoes are sometimes grown in soil samples to test for the presence of root knot nematodes. If you use galling on tomatoes to indicate the presence of *Meloidogyne* spp., avoid Red Cherry or Rutgers cultivars, which produce almost no galling in response to M. *chitwoodi*. Roza, Yellow Pear, Columbia, and Saladmaster do produce gall-like symptoms in response to this root knot species. Other species induce gall formation on all susceptible tomato cultivars.

Differential Host Test. To make decisions about crop rotations and controls, it is important to know which root knot nematode species and races are present. Root knot nematodes isolated from soil samples can be identified to species, but this requires experience and a good microscope; races cannot be identified with a microscope. The differential host test uses a series of indicator plants, each of which is susceptible to a different range of root knot species and races. The test uses the Nugaines cultivar of wheat—a good host for M. *chitwoodi* but a nonhost for M. *hapla*—and the California Wonder cultivar of pepper—a good host for M. *hapla* but a nonhost for M. *chitwoodi*. In areas where M. *hapla* and M. *incognita* occur, but not M. *chitwoodi*, substitute Charleston Grey watermelon for Nugaines wheat. Follow the steps below to carry out the test.

1. Mix a soil sample from the field to be tested (see "Soil Sampling," below), then divide it into four 4- to 6-inch (10- to 15-cm) pots. Plant one to three 3-week-old wheat seedlings in two of the pots and one to three 3-week-old pepper seedlings in the other two pots. Grow the plants at 70° to 80°F (21.1° to 26.7°C) for 55 days. This assumes that a soil extraction has determined there is a high population of root knot nematodes in the soil. If populations are low, plant seedlings of Rutgers tomato (susceptible to all root knot species) in the soil and grow them for 55 days or more to increase the population density before proceeding with the test.
2. Remove the plants from the pots and immerse each root system in fresh water to gently wash out the soil. Remove and discard the tops. Rinse the container

after each root system. Blot each root system with a paper towel to remove excess water. Place each root system in a separate, watertight plastic bag.

3. Prepare a Phloxine B stock solution by adding ½ ounce (14 g) of Phloxine B to 2½ gallons (10 l) of tap water. Label this stock solution and store at room temperature. (Phloxine B can be obtained from sources listed in the suggested reading.)

4. Mix one part of the stock solution with nine parts of tap water. Cover the roots in the watertight plastic bag with this solution, wait 15 minutes, then pour off the liquid. Roots will stain a light pinkish orange; egg masses of root knot nematodes will stain a dark red.

5. Check for egg masses on each stained root system with a magnifying lens or a dissecting microscope, and rate each plant as either a host, if many egg masses are present, or a nonhost, if few or no egg masses are present.

6. Another method for evaluating host status is to extract and count eggs. This can be done instead of evaluating egg masses (steps 3–5) or after egg masses have been rated. Cut roots into segments about 1 inch (2.5 cm) long and place them in 1 pint (500 ml) of a 10% bleach solution in a closed container such as a jar with a screw-top lid. Shake this suspension vigorously for 5 to 10 minutes and pour onto a coarse screen (50 mesh, 300 μ) over a fine-mesh screen (500 mesh, 25 μ). Rinse the root pieces on the screen with water and then discard them. Rinse all debris, which includes nematode eggs, from the fine-mesh screen into a beaker and bring to a known volume, e.g., 200 ml (6.8 oz.) or 8 ounces (236 ml). Stir this egg suspension and transfer a small, carefully measured aliquot, for example, 5 ml (0.17 oz. or 1.0 tsp.), to a counting dish. Count the number of eggs in the dish and determine how many eggs are present in the beaker. If many eggs (thousands) were recovered from the plant, it is a host; if few or no eggs, it is a nonhost.

If wheat plants are hosts, then M. *chitwoodi* is present; if pepper plants are hosts, M. *hapla* is present. Both nematodes may be present in the same soil sample. Watermelon indicates the presence of M. *incognita*, which causes the formation of large, distinctive galls.

If this test indicates M. *chitwoodi* is present, races of M. *chitwoodi* can be distinguished by using Red Cored Chantenay carrot, a good host for race 1 and nonhost for race 2, and Thor alfalfa, a good host for race 2 and nonhost for race 1. Inoculum for alfalfa and carrot must come from a wheat plant to assure that no M. *hapla* is present, because both alfalfa and carrot are hosts for M. *hapla*. To obtain inoculum from wheat, plant wheat (preferably as 3-week-old seedling transplants) into infested soil and grow for 55 days

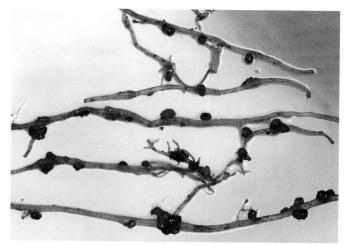

Root knot nematode egg masses on root surfaces are stained dark pink by Phloxine B.

or more to increase the population level. Extract eggs from wheat roots as described above in step 6, determine the total number of eggs in the beaker, and calculate the volume of water that would contain 5,000 eggs. Transplant 3-week-old seedlings of Red Cored Chantenay carrot or Thor alfalfa into potting soil (avoid using peat moss) and pour the volume of water containing 5,000 eggs around each root system as it is being transplanted. Grow for 55 days as described above and extract and count the eggs recovered from the carrot and alfalfa root systems. Divide the number of eggs recovered by 5,000 to determine the level of population increase (R). If R is less than 0.1, the plant is a nonhost; if between 0.1 and 0.9, the plant is a poor host; if between 1 and 2, the plant is a moderate host; if greater than 2, the plant is a good host. If the test indicates carrot is a host, then M. *chitwoodi* race 1 is present. If alfalfa is a host, then M. *chitwoodi* race 2 is present. Both races may be present in the same soil sample.

The Phloxine B stain or the egg extraction method can also be used to test samples of potato roots taken from the field. These tests will indicate the presence of root knot nematodes but will not differentiate between species.

Soil Sampling. A more precise way to monitor for nematodes before a new crop is planted is to take a series of soil samples and submit them to a qualified laboratory for extraction, enumeration, and, if necessary, identification of the nematode species present. Samples should be taken in the fall after harvest or preferably when the previous crop is still in the field; nematodes normally are concentrated in the root zone of plants, and it is easier to find them when the plants are present. It is important to sample while the soil is in good working condition and is neither saturated nor very dry. Contact the laboratory in advance and schedule sampling so that the they can process the samples as soon as possible after receipt. Farm advisors, extension agents, or other authorities

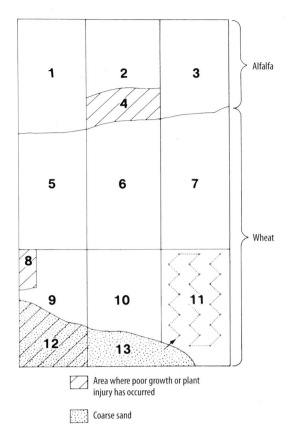

Figure 38. **Before taking soil samples or checking roots for galls, divide the field into areas that reflect any differences in cropping history or soil type. For soil sampling, subdivide these areas into blocks of about 5 acres (2 ha). If there are areas where there has been poor growth, treat them as separate blocks, even if they are smaller than other blocks. Assign each block a number and collect soil from a series of points in each one, following a systematic pattern as shown in block 11.**

can help you find a laboratory equipped for extracting and identifying nematodes from soil samples. Take the following steps, but also contact the lab for any special instructions they may have. For more information on nematode sampling, consult *Sampling for Nematodes in Soil* listed in the suggested reading.

1. Draw a map of the field, showing areas that differ from the rest in soil texture, cropping history, or crop injury (Fig. 38).
2. Divide the area into blocks of 5 acres (2 ha) or less, using a grid pattern. If conditions are uniform throughout the field, apply the grid pattern to the whole field. In general, the smaller the blocks are, the more precise the result of sampling will be.
3. Collect 20 or more subsamples of soil and roots from each block with a soil auger, JMC tube, or Oakfield tube that takes a core 1 inch (2.5 cm) or more in diameter. A shovel may be needed if the ground is too hard or rocky for a soil probe. Take the soil from

the root zone of the crop to a depth of 12 inches (30 cm) or more; take deeper samples from bare ground. Include as many feeder roots as possible. Make sure the soil is moist, but avoid places where soil is wet or compacted. If the surface soil is dry, discard it and include only moist soil.

4. Mix the subsamples from a single block thoroughly in a bucket or bag, then transfer about 1 quart (1 l) into a plastic bag or other moisture-proof container.
5. Label the sample in pencil or indelible marker with the block number. Put the label on the outside of the container, since moisture may ruin labels placed inside. Label all samples with your name and address, the date, location of the field, crop history, the crop present when you took the samples, the crop you intend to plant, and any notes you have of previous crop injury.
6. Keep the samples cool; do not leave them in the sun and do not freeze them. The best storage temperature is 50° to 60°F (10.0° to 15.6°C). Seal the containers so they will not dry out. Samples may be kept in good condition by keeping them in an ice chest until they arrive at the laboratory.
7. Send or deliver the samples to the lab immediately in a cardboard box insulated with newspaper or in an ice chest packed to keep samples from breaking in transit.

It usually takes about 2 weeks for a lab to perform a routine analysis for nematodes. Methods commonly used to extract root knot nematodes from soil recover only juveniles, and the results are usually reported as the number of nematodes found in a standard volume or weight of soil, such as 250 cc or 1 kg. Pay close attention to the units used to report nematode densities when changing or comparing different testing labs. If necessary, ask for help in converting results from different labs to comparable units. The centrifugal–flotation method and the Baermann funnel method are most commonly used for nematode extraction. Each lab report should specify the extraction method used, and it should also indicate the estimated proportion of the nematodes present that are actually extracted. This figure, called the recovery or efficiency rate, is usually from 10% to 30% for root knot juveniles.

Laboratories should identify to genus and list all other plant-parasitic nematodes such as root lesion nematode (*Pratylenchus* spp.), stubby-root nematode (*Paratrichodorus* or *Trichodorus* spp.), and potato rot nematode (*Ditylenchus destructor*). They will also identify, if requested, the species of root knot nematode that are found. Species identification may be determined by microscopic examination or by rearing the juveniles to the adult stage on differentially susceptible hosts, which increases the cost of lab analysis but can be valuable in choosing rotation crops and planning control procedures. Species identification is important when using crop rotation as a management strategy because different crops support reproduction of different nematode species (see Table 16). Race determination of M. chitwoodi by differential host assays also is important because alfalfa supports reproduction of race 2 but not race 1, and therefore may be used to suppress populations of race 1 only.

Besides root knot nematodes, many other kinds of plant-parasitic nematodes are often found in soil samples. Commonly found in potato soils are root lesion nematodes, stubby-root nematodes, stunt nematodes (*Tylenchorhynchus* spp.), pin nematodes (*Paratylenchus* spp.), and spiral nematodes (*Helicotylenchus* spp). Root lesion nematodes and stubby-root nematodes may be of economic importance; none of the others are important on potatoes.

Control

There are no precise guidelines for deciding what number of root knot nematodes found in soil samples will cause unacceptable damage in potatoes. If any M. chitwoodi are found in samples, damage can be expected. Generally, if root knot nematodes are found in fields to be planted to potatoes in the spring and the area has a history of nematode damage to potatoes, soil treatments should be used to reduce the nematode populations. In areas where potatoes are planted in winter and harvested in spring, tubers usually escape significant damage even if root knot nematodes are present. Check with specialists in your area for treatment recommendations based on the results of soil sampling.

Control measures for root knot nematodes include sanitation, crop rotation, green manure cover crops, dry fallow, and nematicide treatments. The choice depends on the effectiveness and cost of these measures, the timing of control decisions relative to season and weather, the value of the crop, and the expected impact of the nematode population under local conditions. Research is underway to develop potato cultivars resistant to root knot nematodes; however, none are commercially available at the present time.

Sanitation. Root knot nematodes do not move far by themselves, so undisturbed infestations spread slowly. However, they can be spread in infected tubers, in soil carried on farm equipment, and in reused irrigation water. A small infestation in one part of a field can easily spread throughout the field when soil is moved in grading. Some species have been widely distributed by the shipment of infested soil and infected tubers.

Follow the sanitation measures listed in the third chapter, "Managing Pests in Potatoes," to avoid spreading infestations of root knot nematodes. The most important points are to use clean seed tubers, avoid moving infested soil or fresh manure or organic matter, and avoid using contaminated water. Using settling ponds reduces the spread of nematodes in surface irrigation water. Leave tubers that remain after harvest exposed to the environment; this helps reduce survival of both nematodes and potato volunteers.

Crop Rotation. Damage from root knot nematodes gets worse and control more difficult where potatoes are grown without rotation or where susceptible crops are grown during rotation between potato crops. Crop rotation does not eliminate infestations because small populations of most root knot nematode species survive in the soil as eggs for up to 2 years in the absence of host crops, and most species also reproduce on a wide range of weeds (see Table 16). Also, almost all rotation crops are hosts for one or more *Meloidogyne* species. However, rotations can reduce the soil population of the dominant nematode species in a given field and make other control measures more effective. Local extension agents, farm advisors, or other experts can help choose suitable rotation crops. Managing weeds and volunteer potatoes during and between rotations is essential for controlling root knot nematodes. In some situations nematicides may be used on rotation crops to reduce nematode populations.

In areas where both M. hapla and M. chitwoodi are present, identification of the species found in soil samples is important in making rotation decisions. Of the crops commonly rotated with potatoes, cereals are nonhosts for M. hapla but good hosts for M. chitwoodi, and they would be the best rotation choice if M. hapla is the dominant species. Alfalfa is a good host for M. hapla and for race 2 of M. chitwoodi, which is widespread in the western states. Asparagus and lima bean are not hosts for either M. chitwoodi or M. hapla. Sweet corn and sudangrass are not hosts for M. hapla, and certain cultivars of these crops are poor hosts or nonhosts for M. chitwoodi. Additional control of harmful nematode species can be achieved by incorporating regrowth of a sudangrass hay crop as a green manure (see below).

In the San Joaquin Valley, southern California, Arizona, and New Mexico, where M. incognita occurs, soil fumigants or nematicides are sometimes applied to control nematodes in rotation crops such as tomato, lettuce, carrot, sugar beet, or cotton. However, nematode populations may build up to damaging levels within one season after a nematicide treatment if susceptible crops are grown. Winter cereals are commonly used for rotation in these areas. Crops such as certain processing tomato cultivars that are resistant to M. incognita, M. javanica, and M. arenaria reduce population levels and reduce the potential for damage to potatoes. Avoid planting winter potatoes following a crop that had severe root knot nematode damage.

Green Manure Cover Crops. Cover crops of rapeseed, oilseed radish, mustard, or sudangrass may reduce root knot nematode populations, as well as populations of root lesion and stubby-root nematodes, when incorporated as green manure. The effect depends on the cover crop and the nematode species. Check with local experts when choosing a cover crop, because some cultivars are suppressive while other cultivars may increase nematode populations. The residues of these crops break down to form compounds toxic to nematodes and certain soilborne fungal pathogens. Sudangrass residues release hydrogen cyanide; residues of mustard family plants release isothiocyanates. Incorporation of cover crops also promotes the buildup of natural enemies in the soil that help reduce populations of harmful nematodes.

Studies have shown significant reductions in damage by Columbia root knot nematode when a winter cover of rapeseed is incorporated before flowering in the spring, or when a late summer cover of sudangrass is incorporated in the fall. Where length of season permits, two mowings of sudangrass for hay can be made before incorporating the regrowth after the second mowing.

Cover crops can be particularly useful before a short-season potato cultivar because the reduced nematode population will have less time to build up to damaging levels before harvest. A short-season cover crop may be advantageous in place of a full-season rotation in areas where water is less available. In potato rotations, rapeseed or sudangrass can be planted in early August following a wheat or sweet corn rotation and incorporated as green manure, sudangrass in fall and rapeseed in fall or the following spring.

Early Harvest. The sooner tubers are harvested the less they are damaged by root knot nematodes. Where winter potatoes are harvested in April or May, damage is usually avoided because populations of M. *incognita* are still low; if tubers are left until July, nematode damage can be severe. Likewise, early potato cultivars are seldom damaged by M. *chitwoodi* or M. *hapla* if tubers are harvested before August. In all growing areas, if root knot nematodes are present, harvest tubers as soon as possible after allowing proper maturation; nematode damage increases the longer tubers are left in the ground, even if nematicide treatments were made before the crop was planted.

Fallow Cultivation. Repeated cultivation of fallow land reduces nematode populations in the upper 12 to 18 inches (30 to 45 cm) of soil by exposing them to heat and to the drying action of the air. Where soil type limits the effectiveness of fumigants, fallow treatments can reduce nematode populations to lower levels than achieved with soil fumigation. In areas where winter potatoes are grown, dry fallow during the summer helps keep nematode populations from increasing. An irrigation during the fallow may improve nematode control by stimulating egg hatch and upward migration. Follow recommended weed control practices to reduce the number of weed hosts and potato volunteers.

Nematicides. Chemicals can reduce populations of root knot nematodes to below damaging levels, but they will not eradicate them. Nematode populations are likely to increase to economically damaging levels within 1 year after treatment if host crops are planted. Nematicides are generally less effective in controlling M. *chitwoodi* because it can multiply more quickly than other root knot species.

All soil fumigants now available are toxic to crops, so a waiting period is required between application and planting. This period is at least 10 days in most cases and may be 3 weeks or more, depending on the material, rate used, and soil conditions. Always observe the waiting period indicated on the label. Fumigating in the fall is usually best because the waiting period before planting is usually not a problem and because spring fumigation may be less effective due to cool, wet soil.

Proper soil moisture and temperature are needed for controlling nematodes with fumigants. Fumigants do not disperse properly in cold, wet soil, so treatments are less effective and the waiting period before planting is longer. Soil that is too dry may allow the fumigant to escape too fast, especially in warm weather. Dry soil does not allow adequate penetration of fumigants applied through sprinklers. Also, more unhatched eggs may be present in dry soil, and nematode eggs are harder to kill than other life stages. Correct moisture levels depend on the fumigant used; follow label recommendations.

Check the soil temperature at the depth of application with a dial thermometer before fumigating. The best temperature for fumigating is 50° to 70°F (10° to 21°C). Do not fumigate if the temperature is below 40°F (4.4°C) or above 80°F (26.7°C). The cooler the soil, the more important that it not be too wet.

Good soil preparation and proper application techniques are also important. Application rates depend on soil type, and procedures may be affected by wind; check label directions. Before applying a soil-injected fumigant, prepare the soil to seedbed condition by deep plowing and discing or rototilling so that the soil is worked up to the desired fumigation depth. Be sure clods are broken up and crop residues are decomposed; fumigants may not penetrate them and nematodes in intact plant parts may not receive a lethal dose.

Soil type can influence the effectiveness of a soil fumigation. Nematode reduction is often less in soil that is high in organic matter. In such cases the level of control achieved with soil fumigation may not be sufficient to prevent serious crop damage when initial nematode populations are high.

Take a soil sample after fumigation and before planting to check the effectiveness of a fumigation. Wait at least 30 days after fumigating to allow dead nematodes to decompose before you sample the treated soil. Depending on the materials and methods used, fumigation may also destroy weeds and kill soilborne plant pathogens such as *Verticillium dahliae*.

Certain nonfumigant nematicides control *M. hapla* but not *M. chitwoodi*. They are frequently used in combination with soil fumigants to control *M. chitwoodi*, especially when populations are high or the nematode is present deep in the soil. Improved control can be achieved when a nonfumigant nematicide is used following incorporation of a green manure cover crop. The effectiveness of a nonfumigant nematicide may depend on the chemical, timing, and application method used to incorporate the material into the soil.

Check the suggested reading in the back of this manual for information about specific chemicals and application procedures that can be used in your area. Also, consult your local extension agent, farm advisor, or other experts for specific nematicide recommendations. Always follow label directions carefully when applying any of these chemicals.

Root Lesion Nematode
Pratylenchus spp.

Several species of root lesion nematode are found throughout the western potato-growing areas, but they usually do not cause economic damage. All feed on roots and most do not attack tubers. Two species, *Pratylenchus penetrans* and *P. neglectus*, can increase the susceptibility of potato plants to Verticillium wilt.

Root lesion nematodes lay their eggs inside the roots of infected plants or in the adjacent soil. The juveniles that emerge from these eggs infect roots that are not suberized. *Pratylenchus penetrans* causes red-brown lesions around the root cortex. These lesions grow together and turn black; under field conditions they may be difficult to see. The lesions are often invaded by soil microorganisms, and yields may be reduced if root damage is extensive. Damage thresholds have been established in other parts of the United States by relating yields to soil populations of *Pratylenchus*. Such thresholds are not presently available for most western growing areas. In the Columbia Basin area, the damage threshold for *P. penetrans* on Russet Burbank appears to be between 500 and 1,000 per ½ pint (250 cc) of soil. These same levels of *P. neglectus* do not affect yield in the absence of Verticillium wilt.

Some species of *Pratylenchus* cause serious damage on potato tubers, but they are not known to occur in the western United States. Tubers can be attacked by *P. neglectus*, but economic damage to tubers is not generally a problem with this species in the western states.

Crop rotation is not an effective means of controlling root lesion nematodes because they have a wide host range. They are particularly favored by grain crops and alfalfa. Populations can be reduced by green manure cover crops, as discussed above for root knot nematodes.

Follow the sanitation recommendations for root knot nematodes to prevent the spread of root lesion nematodes.

Fumigation for root knot nematodes controls root lesion nematodes as well, but fumigation usually is not recommended to control *Pratylenchus* alone. If root lesion nematodes are causing a problem, nonfumigant nematicides usually control them. Consult your extension agent, farm advisor, or other experts for specific recommendations.

Stubby-Root Nematode
Paratrichodorus (Trichodorus) spp.

Though widespread in the western states, stubby-root nematodes are of economic concern only in certain areas of California, Colorado, Idaho, Oregon, and Washington, where they are vectors for *Tobacco rattle virus* (TRV), the virus that causes corky ringspot of potato tubers. Stubby-root nematodes are confined to coarse, sandy soils, in which they can live as deep as 40 inches (1 m) or more. They attack the roots of a wide range of plants, feeding on root surfaces and causing the formation of numerous stubby roots. Root damage does not affect potatoes significantly; the nematodes only become a problem when they transmit TRV.

PHILIP B. HAMM

The root lesion nematode *Pratylenchus penetrans* causes red-brown lesions around the root cortex.

Potato rot nematode causes dark brown, honeycombed, granular rot. Rotted tissue becomes wet and soft if invaded by bacteria.

Female potato cyst nematodes form egg-containing cysts, which are $^1/_{50}$ to $^1/_{30}$ inch (0.5 to 0.8 mm) in diameter and contain up to 500 eggs, on the surfaces of host plant roots. The potato cyst nematode does not occur in the western states.

A large number of crops and some weeds are hosts of TRV. Stubby-root nematodes acquire the virus when feeding on infected roots, and they transmit the virus when they move to a new host. Tubers infected with TRV develop corky ringspot (discussed and illustrated in the fifth chapter, "Diseases").

Avoid moving soil from fields where corky ringspot is found; if possible, do not plant potatoes in these fields unless they are properly treated. In some states, fields where corky ringspot has been found are permanently prohibited from growing certified seed potatoes.

Stubby-root nematode can be controlled effectively by fumigants if they are applied properly to achieve deep penetration and adequate diffusion. Application in irrigation water (chemigation) is not effective. Repeated application of nonfumigant nematicide can protect tubers from nematode infection and virus transmission. Fumigation followed by nonfumigant nematicide is more effective than using either fumigant or nonfumigant alone.

If you want to determine whether stubby-root nematodes are present in fall soil samples, be sure to sample to a depth of at least 24 inches (60 cm), because these nematodes move deep into the soil at that time of year. Populations as low as 3 per half-pint (250 cc) of soil may be able to transmit enough TRV to cause symptoms that will result in crop rejection.

Potato Rot Nematode
Ditylenchus destructor

In the West the potato rot nematode has been reported to cause damage to potatoes only in a few isolated areas of Idaho. The nematode overwinters as eggs and survives on a large number of plants, including grains, tomatoes, carrots, bulbous lilies, and several weed species. *Ditylenchus* is more frequently a problem in the production of bulbous iris or dahlia bulbs than in potatoes.

Juveniles hatch from eggs and attack stolons and tubers, entering tubers through eyes or lenticels. Symptoms appear only on tubers, first as pinhole-sized lesions that may be confused with wireworm damage. The nematodes can be seen just below the skin as pearly white spots. Lesions usually develop in storage; the nematodes are active at temperatures as low as 40°F (4.4°C). Cavities develop around the nematodes, and the tuber surface becomes dry and cracked. The rot usually takes on a chalky, honeycombed appearance. If the chalkiness is not apparent, symptoms may be confused with scab or tuber decay caused by ring rot or late blight, but nematodes are easily seen in rotted tissue examined with a dissecting microscope.

Ditylenchus may be controlled by soil fumigation. Avoid planting potatoes where the nematode occurs. Avoid moving infested soil or infected tubers. Do not store tubers if symptoms are seen during harvest.

Potato Cyst Nematode
Globodera rostochiensis

The potato cyst nematode, also called the golden nematode, occurs in some potato-growing areas of Canada and the eastern United States. Strict quarantines and seed certification requirements have thus far prevented it from spreading to other potato-growing areas on infested tubers. Although potato cyst nematode does not occur in the western states, potato growers and pest control advisers should be prepared to recognize its symptoms in case it is introduced. If potato cyst nematode is introduced, prompt control action may prevent it from becoming established or spreading. Once established, this nematode is almost impossible to eradicate, because it forms egg-containing cysts that can survive for at least 20 years in the absence of a host.

Weeds

Weeds reduce yields by competing with potatoes for light, water, and nutrients. Weeds may also be hosts for insects, pathogens, or nematodes that attack potatoes. Dodder, a parasitic weed, injures potato plants directly, and the rhizomes of nutsedge and quackgrass may penetrate tubers, causing losses in quality. Heavy weed infestations reduce harvesting efficiency, increase mechanical damage to tubers, and increase the number of tubers left in the ground or lost over deviner chains.

Management Guidelines

An effective weed management program takes into account the primary weed problems, crop rotation, cultivation, available herbicides, and the competitive ability of the potato crop. Competition from early-season weeds reduces potato yields if the weeds are not controlled within 4 to 6 weeks after potatoes emerge. Weeds that emerge after vines have covered the rows usually do not compete with the potato crop; however, they may reduce yields by interfering with harvest and can produce seed that infest subsequent crops. Weeds frequently become more serious if potatoes are set back by adverse conditions early in the season. Weed problems can be reduced by establishing a vigorous stand of potatoes. Vigorous indeterminate cultivars such as Russet Burbank have fewer weed problems than less vigorous determinate cultivars such as Russet Norkotah. For this reason, mid- and late-season weed control is more likely to be needed on determinate cultivars. In all cases, weed management will be easier when you prepare fields properly, plant good-quality seed tubers, use spacings that give an adequate plant stand, maintain adequate soil moisture and fertility, and control diseases, insects, and nematodes.

Field Selection and Crop Rotation

Efficient weed management requires matching the right crop with the right field. You must know which weeds will be present if you are to make this choice. Survey the weeds in the field during the previous crop, then check to see if herbicides will control the major weeds present. Be sure that conditions will permit proper application of the necessary herbicides. Careful field preparation and irrigation control are required for herbicides that are applied through sprinklers and for some other types of applications as well. Some herbicides may cause injury if applied under poor growing conditions. Fields with major variations in soil type may require changes in herbicide rates or avoidance of some herbicides.

Crop rotations help control difficult weed problems because they allow a variety of weed control methods. Different planting times and cultural practices required for rotation crops may eliminate conditions favoring particular weeds; different herbicides and tillage practices are usually available in alternate crops. Choose rotation crops that compete well with problem weeds and allow effective alternative herbicides, then follow active weed control programs in those crops. Grains are valuable in rotations; herbicides that cannot be used in potatoes can be used for controlling problem weeds in the grain crop and after harvest. A winter grain or winter canola crop can reduce populations of common summer annual weeds such as nightshades and wild oat. Corn competes well with weeds, tolerates a variety of herbicides, and may also help reduce populations of pathogens and harmful nematodes. Control of troublesome perennial weeds may be achieved by discing and watering after a grain harvest to encourage weed growth in the fall, then applying a translocated (systemic) herbicide to the regrowth of the perennials. Cotton rotations allow alternative herbicides. Alfalfa may be a useful rotation crop because it shades out a number of annual weeds common in potatoes, and repeated mowing of alfalfa helps reduce infestations of some troublesome weeds. In addition, the competitive canopy of alfalfa helps reduce yellow nutsedge infestations. However, the susceptibility of alfalfa and cotton to certain root knot nematodes may diminish their usefulness. Other potato family crops such as tomato, pepper, and eggplant are not desirable in rotations because the available herbicides are similar to those used in potatoes, and these crops are often hosts of pathogens, nematodes, or insect pests that attack potatoes.

Check with your local extension agent, farm advisor, or other experts to find out which herbicides are registered for use in the rotation crops suited to your area, then choose a combination of crop and herbicide that controls the weeds present. Check the labels to make sure herbicide residues will not damage subsequent crops. Avoid repeated applications of the same herbicide or herbicides with similar modes of action. The frequent use of herbicides with the same mode of action encourages buildup of weeds not controlled by those herbicides and the possible buildup of herbicide-degrading microbes in the soil or herbicide-resistant weed populations.

Monitoring

To plan a weed management program, you must know what kinds of weeds are present, which ones are most abundant, and whether their abundance is changing. Most of the herbicides used in potatoes are effective only on germinating or very young weeds, so it is essential to know what the target weeds are before they emerge. Normally this information comes from routine weed surveys made in previous seasons. Keep separate records for all fields; this helps you select the correct controls and detect changes in weed populations that may require special attention.

Monitoring is useful for deciding if additional herbicide applications are necessary and may help reduce the amount of herbicide needed for weed control. After applying an herbicide such as EPTC, which is easily leached from the soil, watch for weed emergence in areas such as the centers of center-pivot sprinklers or along the lateral lines of solid set sprinklers, where most leaching occurs. If using metribuzin, you can reduce the likelihood of injury by using lower rates, then monitoring your fields and applying metribuzin again if weeds emerge.

Surveys. Regularly survey your fields to look for changes in weed populations and to check the effectiveness of weed control measures. Surveys can be made on foot or from field equipment during routine field operations, and can be made when monitoring for other pests. If possible, make your first survey while the previous crop is still in the ground. Repeat surveys during each growing season to check on the effectiveness of weed control measures. Make surveys once each week during the early part of the potato season.

Make a map for each field during the first few weeks after crop emergence. Use the maps to evaluate the effectiveness of preplant or preemergence controls and the need for cultivations or postemergence herbicides.

Cover the entire field when making a weed survey and rate the degree of infestation for each weed species. Check fencerows and ditchbanks and record these separately on the field map. Pay special attention to perennials and dodder infestations and mark where they occur on the field map. This allows you to recheck these areas easily each season and apply special controls during fallow or rotations. Make a final weed survey after vines begin to senesce or during harvest.

Weed survey forms are helpful for keeping records; a sample is shown in Figure 39. Whether you use a printed form or not, be sure to take written notes in the field and keep them as part of the permanent field history. Weed survey information collected over several years is extremely valuable in identifying changes in weed populations and in planning herbicide and rotation programs.

Soil Tests. If the field is plowed before you can do a weed survey, use a simple soil test to identify at least some of the weeds present. The test does not work for certain weeds, such as shepherd's-purse, and results are less reliable than those of a survey, but they are better than no information.

Collect samples of the top 2 to 3 inches (5 to 7.5 cm) of soil at random from at least 10 to 20 places in each field. Mix the samples from each area and put the soil in a greenhouse flat or similar container. Moisten the soil with a solution of 100 parts per million (0.01%) gibberellic acid (available commercially) and keep the soil indoors where it is warm and sunny. Use the photos in this manual and in the weed identification handbooks listed in the suggested reading to identify

WEED INFESTATION RECORD

FIELD LOCATION	CROP
PLANTING DATE	SCOUTING DATE
PREVIOUS CROP	HERBICIDE/DATE APPLIED
HERBICIDES IN PREVIOUS CROP	COMMENTS

	Dodder			Field pennycress			Sowthistle
PERENNIALS				Flixweed			Sunflower
	Canada thistle			Knotweed			Tansymustard
	Field bindweed			Kochia			Tumble mustard
	Nutsedge			Lambsquarters			Wild buckwheat
	Quackgrass			London rocket			
				Mallow		**ANNUAL GRASSES**	
BROADLEAF ANNUALS				Mustards			Barnyardgrass
	Black nightshade			Nettleleaf goosefoot			Foxtail
	Cutleaf nightshade			Pigweed			Sandbur
	Hairy nightshade			Purslane			Volunteer grain
	Groundcherries			Russian thistle			Wild oat
	Cocklebur			Shepherd's-purse			Witchgrass

Figure 39. Example of a weed monitoring sheet. To indicate the level of infestation, use a scale from 1 to 5 in which 1 = very few weeds, 2 = light infestation, 3 = moderate infestation, 4 = heavy infestation, and 5 = very heavy infestation. Alternatively, you can rate the infestation level as L = light, M = moderate, H = heavy.

weed seedlings that emerge. Call your local extension agent, farm advisor, or other experts if you need help.

Herbicide Residue Test. Certain herbicides used in rotation crops may leave residues harmful to potatoes. Among those that can leave harmful residues are atrazine, chlorsulfuron, clomazone, clopyralid, dicamba, diuron, hexazinone, imazaquin, imazethapyr, linuron, mesotrione, picloram, simazine, and terbacil. Consult the labels of herbicides that have been or may be used in rotation crops for restrictions regarding the planting of potatoes following their use. If potentially harmful herbicides have been used, you may want to test the soil to see if residues have dissipated before planting a potato crop. Take samples from several different places in the field. At each sample location, dig soil down at least 1 foot (0.3 m), mix it thoroughly, and take enough to fill each of three to five soil buckets or pots with about 5 to 10 pounds (2.2 to 4.5 kg) of soil. As a check, take at least one sample from a location you are sure is free of herbicide residues. Plant one or two seed pieces in each of the buckets or pots, apply some starter fertilizer, and grow the plants until they are at least 4 inches (10 cm) high. Look for injury symptoms by comparing the test samples with the checks. (See "Herbicide Injury" in the sixth chapter, "Physiological Disorders," for examples of herbicide injury symptoms.) The results of a soil test should never preclude the label restrictions of herbicides that have been used in rotation crops.

Cultivation

During field preparation and hilling, cultivation is an effective way to manage annual weeds early in the season. Cultivation is often the only management option available for perennial weeds after potatoes have emerged; it is essential for managing annual weeds that are not controlled by available herbicides. Cultivations during hilling operations enhance weed control by uprooting and burying seedlings. Use a rolling cultivator behind hilling blades to increase the uprooting of early-emerging weeds during hilling operations. The weed control provided by cultivations may eliminate the need for herbicide applications in some situations. However, the general recommendation for Russet Burbank is to keep cultivation and hilling operations to a minimum. When weed populations are low and a potato cultivar that competes well is grown, cultivation when weeds are small and potato plants are 4 to 6 inches (10 to 15 cm) tall can provide economical weed control.

Additional cultivation can be used about 30 to 40 days after potato emergence to control later-emerging weeds. If cultivation controls weeds until potato row closure, herbicides may not be needed. However, solid set sprinklers may restrict the number of cultivations possible in some areas. Cultivations in addition to normal hilling generally are not recommended in most areas because they can compact soil, damage potato roots and shoots, cause loss of soil moisture, and spread pathogens. Multiple cultivations reduce yields.

Sanitation

The spread of problem weeds into potato fields is reduced by controlling weeds along field edges. Common weeds that may be present near fields are also hosts for insect and disease pests that may move into potatoes.

The most common way weeds are introduced to a field is on contaminated tillage and harvest equipment. Use compressed air, steam, or water to clean equipment between fields, especially when leaving fields with infestations of perennial weeds. Work fields with fewer weed problems first before moving to more heavily infested fields.

Where canals are used for irrigation water, install screens at headgates and pumps to exclude weed seeds and propagules. Use certified seed for rotation crops to reduce the likelihood of introducing weeds. Properly compost manure before applying it to fields to reduce the number of weed seeds that may be introduced.

Herbicides

Herbicides can play an important role in managing weeds in potatoes. Using them in combination with cultural practices such as crop rotation and cultivation increases their effectiveness.

Soil-applied herbicides, which are usually mixed mechanically or irrigated into the soil before weeds emerge, kill germinating weed seeds or emerging weed seedlings. Foliar-applied herbicides are sprayed onto the foliage of weeds after they have emerged, usually while they are small and actively growing. Glyphosate is a foliar-applied translocated herbicide; when applied to the foliage it moves throughout the plant, killing both aboveground and belowground parts. Translocated herbicides are most effective when applied to actively growing plants. Glyphosate kills potato plants as well as weeds, so when used in potato fields, it must be applied to weeds before potato emergence.

With the exception of diquat, glyphosate, and paraquat, most herbicides used in potatoes are selective, killing some weeds but not others; most kill only germinating or newly emerged weeds. Choose herbicides that control the important weed species in your field, and plan to kill weeds before they grow beyond the seedling stage. Some herbicides may

be used in combination to increase the spectrum of weeds controlled. Information on the susceptibility of weed species to available herbicides can be found in *UC IPM Pest Management Guidelines: Potato* and *Pacific Northwest Weed Management Handbook*, listed under "Weeds" in the suggested reading. Check these resources for specific recommendations on herbicide use. Such recommendations are not made in this book because registrations change frequently and may differ between states.

Depending on the herbicide used and the weeds to be controlled, the time of herbicide application may be preplant, preemergence (before potatoes emerge), or postemergence (after potatoes emerge). Special precautions may be necessary with preemergence or postemergence application of some herbicides to avoid injuring potato sprouts or vines.

Details of a specific weed control program depend on the weed species present, soil type, cultural practices, cultivar grown, rotation crops, and available herbicides. Two or three herbicide applications are usually made during the potato season. A preplant soil-applied herbicide may be mixed into the soil during field preparation and broadcast fertilizer application. Depending on the weeds in question, preemergence herbicides may be applied in some areas about 3 weeks after planting. Usually a combination of two or three soil-applied herbicides is mixed into the soil during or just after hilling. Residual control usually can be extended later into the season by hilling 2 to 3 weeks after planting and applying herbicides immediately, or by hilling at planting and waiting until just before potatoes emerge to apply herbicides. Another alternative is to form hills just before potatoes emerge and apply postemergence herbicides after potatoes and weeds have emerged and before vines close. Vine-killing agents are usually contact herbicides.

Proper application techniques increase herbicide effectiveness and decrease the likelihood of crop injury. Determine exactly how the herbicides you choose are to be applied and plan every detail into your cultural program. Some can be applied through overhead sprinklers; this requires a well-designed system with accurate metering, flow-controlled nozzles where land is not level, and check valves to prevent contamination of the water supply. Check with your local extension agent, farm advisor, or other experts before applying herbicides in this way. Never apply herbicides by sprinkler irrigation if the wind speed is over 10 miles per hour (16 km/h). The herbicide EPTC can also be applied in furrow irrigation water.

Proper herbicide rates depend on the weeds to be controlled and on the soil in your field. Higher rates of some herbicides are required on soils high in organic matter. Consult your local extension agent, farm advisor, or other experts for the rates that work best in your area. Some herbicides may damage certain cultivars. If you do not know the sensitivity of a cultivar, test the application on a few plants first. Find out whether subsequent crops will be affected by the herbicides you use and whether special efforts such as deep plowing will be needed to protect them from herbicide residues. ALWAYS FOLLOW LABEL DIRECTIONS CAREFULLY, and be aware that registrations and restrictions change.

Herbicide Resistance. Repeated use of the same herbicide or herbicides with the same mode of action may favor the buildup of weed populations that are no longer controlled effectively by that herbicide. In an IPM program, a variety of weed management techniques are used, which decreases the risk of selecting herbicide-resistant weeds. Follow these guidelines to reduce the likelihood that resistant weed populations will develop.
- Rotate herbicides and crops.
- Use cultivation and other cultural practices to supplement the use of herbicides.
- Avoid sequential applications of the same herbicide.
- Use tank mixes of herbicides with different modes of action.
- Use tank mixes that include herbicides with overlapping weed spectra so herbicides with different modes of action are used to control the same weed.
- Use proper application techniques to ensure herbicides are as effective as possible.
- Monitor fields before and after herbicide applications to assess effectiveness and make the best decisions about subsequent control strategies.
- Keep records for each field of herbicides used and weeds present so you can rotate to herbicides with different modes of action and keep track of how well weeds are controlled.

Major Weed Species in Potatoes

The weeds that infest potatoes can be grouped biologically into four categories, each of which may require special control practices: perennial weeds; a parasitic weed (dodder) that feeds directly on potato plants; annual broadleaf weeds; and annual grasses. Annuals complete their life cycle of germination, growth, flowering, and seed production within 1 year. Winter annuals germinate in the fall, grow through the winter, flower in late winter and early spring, produce seed in the spring, and die by early summer. Summer annuals germinate in late winter, spring, or early summer and produce seed in summer or fall; they die in fall or early winter. Perennials live for 3 or more years, usually dying back in the winter and regrowing from underground vegetative structures. Although perennials do not always produce seed during their first year of growth, they produce seed each year once established.

Vegetative plant parts used in identifying weed species are illustrated in Figure 40. Geographic location, growing conditions, and cultural practices determine which weed species may be

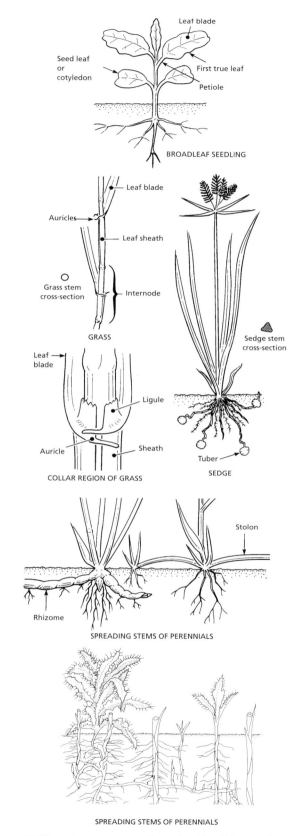

Figure 40. Vegetative parts commonly used in identifying weeds.

Table. 17. Weed Emergence Dates for Potato-growing Areas in the Western States.

Weeds	Weed emergence dates in growing area (see Fig. 41)									
	1	2	3	4	5	6	7	8	9	10
barnyardgrass	Apr–May	Apr–Aug	May	Apr–Jun	Jun–Jul	Apr–Aug	Apr–Aug	Feb–Mar	Mar	May
Canada thistle	Mar–f¹	Mar–Oct	May	Mar–f	Apr–May	—²	Mar–f	—	—	—
cocklebur	May	May–Aug	May	—	May–Jun	May–f	May–f	Mar–Apr	Mar–Apr	Jul
dodder	Apr–f	May–Jun	Jun	Apr–f	Apr–May	May–f	—	Apr–May	—	May
field bindweed	Mar–f	Apr–Sep	May	Mar–f	May–Jul	Apr–f	Apr–f	Mar	Mar	May
foxtails	May–f	Apr–Aug	May	May–f	Apr–Jun	Mar–f	Apr–f	—	Mar	Mar
groundcherries	—	May–Aug	—	—	—	Jun–f	—	Mar–Apr	May	May
kochia	Mar–f	Apr–Aug	Jun	—	Apr–May	Mar–f	Mar–f	—	—	—
lambsquarters	Mar–f	Apr–Aug	Apr	Apr–f	May–Jun	May–Aug	Mar–f	Sep–Oct	Feb	May
london rocket	—	Aug–Mar	Apr	—	—	—	—	Oct	Aug–Mar	Apr
mallow	Mar	Apr–May	May	Mar–Apr	Apr–May	May–f	—	Oct	Sep–Feb	Mar
mustards	all year	Aug–May	Apr	all year	May–Jun	Jan–Mar	Jan–Mar	Oct	Aug–Mar	Mar
nettle	—	—	—	—	—	Jun–Jul	—	—	Nov–Feb	all year
nightshades	Apr–f	Apr–Sep	May	Apr–f	Jun–Jul	May–f	—	—	Feb–Aug	May
nutsedge	Apr–May	Apr–Sep	—	Apr–May	—	—	—	Mar–Apr	Mar	May
pigweeds	Apr–May	Apr–Aug	Apr	May	May–Jun	Mar–Aug	Mar	Feb–Mar	Mar–Apr	Apr
purslane	May–Jul	Jun–Jul	Jun	Apr–Jul	Jun–Jul	Apr–Jul	—	Mar–Apr	Apr–May	May
quackgrass	Mar–f	Mar–Oct	Apr	all year	Apr–May	—	May	—	—	—
Russian thistle	Feb–Apr	May–Aug	Jun	—	May–Jun	Apr–Jul	May	Mar,Sep	Feb	Apr
sandbur	May	May	—	—	Apr–Jun	Jun–Aug	May	Feb–Mar	—	—
shepherd's-purse	Feb–Mar Sep–Oct	Sep–May	Apr	Sep–Jun	Jun	Feb–Mar	—	Oct	Aug–Mar	Mar
sowthistle	Mar–Apr	Sep–Jun	May	all year	May	—	—	Oct	all year	Mar
sunflower	May–Sep	May–Aug	—	—	Jun	May–Jun	May–Jun	Mar–Apr	—	—
wild oat	Mar	Mar–Jul	Apr	Sep–Jun	Apr–May	—	—	Oct	—	Mar
witchgrass	May–Aug	May–Aug	—	May–Aug	Jun	—	—	—	—	—

1. f = frost
2. Dash indicates weed is not important in the area.

problems in a particular growing area. Weeds that occur in potatoes to a significant extent in one or more of the potato-growing areas of the western states are discussed in this section. Emergence dates of weed species important in the growing areas identified in Figure 41 are listed in Table 17.

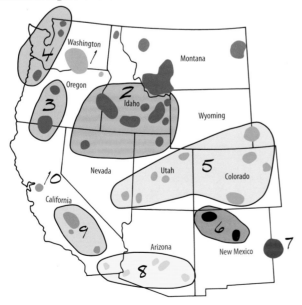

Figure 41. Potato-growing areas of the western states that have similar weed-emergence dates. Estimates of weed emergence for these areas are listed in Table 17.

Perennial Weeds

Perennial weeds are often the most difficult to control because they have persistent rhizomes, roots, or tubers that survive when aboveground parts of the plants are killed. Avoid fields with heavy infestations whenever possible. Control perennials before planting potatoes; once they establish in a growing crop, little can be done to control them except roguing and cultivating. Control seedlings whenever possible to prevent plants from getting established. The basic strategy for established plants is to destroy as much of the vegetation as possible, then prevent the regrowth of the green parts. Useful methods include planting rotation crops in which alternative cultural practices and herbicides can be used, using herbicides during the season that prevent or retard top growth of perennials, using foliar-applied translocated herbicides when crops are not present, and fumigating the soil. If the growth of green vegetative parts can be prevented, the energy stored in underground parts will eventually be exhausted and the infestation will die out. Serious perennial weed infestations require persistent control programs for several years.

Apply a selective translocated herbicide to new weed foliage during a rotation, or a nonselective translocated herbicide after harvest. Apply herbicides to new leaves that are actively growing but not under stress.

Soil fumigation controls quackgrass and partially controls other perennial weeds. Field bindweed and Canada thistle are usually only set back by fumigation because their perennial roots penetrate so deeply. Generally, soil fumigation is economical for weed control only in small, heavily infested areas or where it is also necessary to control nematodes or Verticillium wilt.

Because rhizomes, roots, and tubers of perennial weeds are readily moved with soil and water, sanitation is essential to prevent infestations from spreading. Destroy isolated infestations by hand or with herbicides. Avoid cultivating infested fields when the soil is moist unless herbicides are to be used; otherwise, large numbers of new plants may start from chopped up rhizomes or roots. Clean equipment to prevent moving infested soil, and put screens in irrigation ditches to prevent moving viable plant parts in water.

Canada Thistle *Cirsium arvense*

Found in most of the western states, Canada thistle can become a serious weed pest wherever it occurs because its extensive perennial root system makes it difficult to control. Canada thistle can be a problem in all potato-growing areas except Arizona, central and southern California, and central and northwestern New Mexico.

Seed production by Canada thistle is sometimes limited because male and female flowers are formed on separate plants, and entire patches of Canada thistle may be only male or female. Seed are spread by wind and may be transported as a contaminant in commercial seed; when buried they remain viable for many years. Seedlings begin forming perennial rootstock when they are 3 weeks old or have five leaves. The roots act as storage organs; if chopped up during cultivation, each piece containing a bud can regenerate a new plant. Shoots from Canada thistle roots begin generating food reserves immediately in the summer or fall; in the spring they generate food reserves when about 1 foot (0.3 m) tall.

Canada thistle control requires cultivation and herbicides in both potatoes and rotation crops. The objective is to set back top growth before it can regenerate food reserves or produce seed. Sanitation is needed to prevent spread to new areas and to eliminate the weed from noncrop locations. Most herbicides that can be used in potatoes do not control Canada thistle. Some available materials suppress top growth. Single applications of a nonselective translocated herbicide are effective if applied when thistle plants are at late bud or early bloom stage, or if applied to small Canada thistle plants in the fall when soil moisture is adequate; this type of material must be used in noncrop areas or when crop plants are not present. Mowing alfalfa hay or using selective translocated herbicides in small grains or corn reduces Canada thistle infestations.

Plan rotations to allow time for a fall herbicide treatment before the final killing frost. After mowing alfalfa or discing grain stubble, irrigate to encourage regrowth of the thistles. Wait at least 3 weeks or, if possible, until after the first light frost, then apply a translocated herbicide; this treatment is effective as long as thistle plants are still green, even after several hard frosts. If potatoes are to be planted the following spring, avoid using high rates of herbicides that may leave harmful residues. This program also controls field bindweed infestations.

A. Canada thistle flowers form purple or occasionally white heads about ½ inch (12 mm) in diameter.

B. Canada thistle root systems may be 20 feet (6 m) deep and extend 15 feet (4.5 m) horizontally.

Quackgrass, *Elytrigia repens*

Quackgrass, also known as couchgrass or wiregrass, is found throughout the northern half of the United States. Quackgrass rhizomes may penetrate potato tubers, reducing their

A. Canada thistle flowers

B. Canada thistle root system

C. Quackgrass rhizomes

D. Quackgrass collar

E. Quackgrass inflorescence

quality. If rhizomes are chopped up, each piece can give rise to a new plant; quackgrass rhizomes are easily spread from field to field on contaminated cultivation equipment. Plants produced from rhizomes emerge 1 to 3 weeks earlier than those produced from seed. New plants begin producing rhizomes when they have three or four leaves.

You can use a nonselective translocated herbicide to control quackgrass before potatoes are planted or in noncrop areas. Apply herbicide to actively growing quackgrass that is at least 8 to 10 inches (20 to 25 cm) tall. For complete control, repeated applications may be necessary. Wait at least 5 to 7 days before preparing seedbeds.

Soil fumigation controls quackgrass but is usually economical only for small infestations, if the quackgrass infestation is severe, or if it is necessary to control nematodes or Verticillium wilt.

Cultivation and herbicide applications may need to be repeated to eliminate quackgrass problems. Observe strict sanitation to avoid spreading or reintroducing quackgrass on contaminated equipment or in irrigation water.

C.   Quackgrass rhizomes are straw colored, with a scaly appearance and sharply pointed ends that may penetrate potato tubers. Most rhizomes are in the upper 6 inches (15 cm) of soil; they extend up to 8 inches (20 cm) deep and 3 to 5 feet (1 to 1.5 m) laterally.

D.   The collar region at the base of each quackgrass leaf has a pair of auricles and a ragged fringe.

E.   Quackgrass seed are formed in a head that resembles a slender, flattened head of wheat. Seed are not a major means of quackgrass survival.

Nutsedges, *Cyperus* spp.

Yellow nutsedge, *Cyperus esculentus*, also called yellow nutgrass, is a serious weed pest of potatoes in eastern Oregon and in the San Joaquin Valley of California, where it is widespread in other crops such as cotton and tomatoes. Nutsedges can become serious in areas of Arizona, Idaho, New Mexico, Oregon, and Washington. They can rapidly spread to new areas because their rhizomes and tubers, or "nutlets," are easily carried in soil on farm equipment. Nutsedge rhizomes may penetrate potato tubers, particularly when soil moisture is low. The greatest damage is likely if nutsedge growth occurs between vinekill and harvest.

Nutsedges can reproduce from seed, but tubers are the main means of reproduction and spread. Plants produce tubers on rhizomes, the majority of tubers being produced in the upper 8 inches (20 cm) of soil, but some are produced deeper. The tubers remain viable for several years in dry soil. When conditions are favorable, new aboveground plants develop from tubers, and the tubers grow a new system of roots and rhizomes. New rhizomes in turn produce additional shoots or new tubers; environmental conditions determine whether shoots or tubers are produced. If new sprouts are killed back, a tuber can sprout up to five more times before exhausting its energy reserves. Long-term control of nutsedge requires preventing seedling plants from growing beyond the five- to six-leaf stage, when they start producing new tubers.

Yellow nutsedge does not tolerate shade; once potato vines have closed over, further nutsedge growth is usually suppressed. In early-season areas, herbicides may not be necessary if potato vines cover the ground before nutsedges begin to emerge. Where nutsedges emerge before vines close, herbicides can prevent nutsedge growth until vines have closed.

Thoroughly apply vine-killing agents to suppress nutsedge growth after vine death and to help prevent tuber damage. Maintaining soil moisture after vine death may also help prevent damage to potato tubers by nutsedge rhizomes. Several herbicides can control or suppress nutsedges in rotation crops; some very effective herbicides are available for grains or alfalfa. However, it may not be possible to eradicate nutsedge even when weed control is used during fallow periods.

Purple nutsedge, or purple nutgrass, *C. rotundus*, is found in the southern areas where yellow nutsedge occurs. It is not as widespread as yellow nutsedge and tends to be localized in wet areas of fields. Purple nutsedge tubers are more susceptible to drying than those of yellow nutsedge; repeated cultivation of dry soil can reduce or eliminate infestations. Purple nutsedge is less susceptible to some herbicides than yellow nutsedge.

F.   Young nutsedge plants are grasslike, but leaves are thicker and stiffer than most grasses. The leaves are V shaped in cross-section and arranged in a spiral at the base.

G.   Nutsedge grows to a height of 1 to 2 feet (0.3 to 0.6 m). Leaves and stems have a waxy appearance. Flowering stalks are triangular in cross-section and are usually no longer than the basal leaves in yellow nutsedge (shown here). The flowering stalks of purple nutsedge are usually longer than the basal leaves.

H.   Tubers of yellow nutsedge are smooth, about ¾ inch (19 mm) in diameter, and have a pleasant, almondlike flavor. They are formed singly on rhizomes, and young tubers have loose outer scales that later drop off. The tubers of purple nutsedge (not shown) are formed in chains along rhizomes. They are scaly and reddish with a bitter flavor.

Field Bindweed, *Convolvulus arvensis*

Field bindweed, also called wild or perennial morningglory, is severe in some potato fields in several of the western states. Heavy infestations smother potato plants and interfere with harvest. It is a serious weed pest of potatoes in the Clovis, New Mexico, area.

Field bindweed spreads from an extensive root system as well as from seed. Roots may penetrate to a depth of 25 feet (7.6 m) or more and extend laterally several feet. When

F. Yellow nutsedge plant

G. Yellow nutsedge flower stalk

H. Yellow nutsedge tubers

I. Field bindweed seedling

J. Field bindweed flowers

roots are chopped up by cultivation in moist soil, the individual pieces can generate new plants. Follow careful sanitation procedures to prevent moving bindweed on contaminated cultivation equipment. Field bindweed seed remain viable for up to 50 years, some germinating to produce new seedlings each year. Preplant or preemergence herbicide applications prevent bindweed seedlings from becoming established during the potato growing season. Once bindweed infestations are established, they are extremely difficult to eliminate.

Some translocated herbicides are useful for control of field bindweed, although they may not kill plants in a single application. For best results with a nonselective material, disc the infested area in the fall 4 to 8 weeks before applying the herbicide. Irrigate to stimulate regrowth of the bindweed, and follow with the herbicide application. Disc the field again after 1 to 3 weeks. If applying a nonselective material to bindweed in the summer, treat when flowers are present on more than half the length of the stem. Herbicide can be applied before potatoes emerge and can be repeated in the fall after harvest; it can also be applied in the fall following a grain rotation when potatoes will be planted in the spring. Applying selective translocated herbicide in grain rotations sets back field bindweed. Repeated treatments over several seasons are necessary to control bindweed.

I.   Field bindweed seedlings have seed leaves that are nearly square with a notch at the tip. The first true leaves are shaped like a heart or an arrowhead. Petioles are flattened and grooved on the upper surface. Plants sprouting from roots (not shown) lack seed leaves.

J.   Prominent white or pink flowers make field bindweed easy to identify. Severe infestations can make harvesting difficult.

Parasitic Weed

Dodder, *Cuscuta* spp.

A parasite that attacks many crops, weeds, and native plants, dodder is a problem in the lower Snake River Valley and in some potato fields of the Columbia Basin and Klamath Basin, but it is not a widespread pest of potatoes. However, its wide host range and the long life of its dormant seed make dodder hard to control and nearly impossible to eradicate.

Dodder lacks chlorophyll and can grow only by penetrating host plant tissues to obtain water and nutrients. Seedlings must attach to a suitable host within a few days of germination or they die. Young seedlings twine around the host plant and produce suckers, called haustoria, that penetrate host tissues. After attachment, the part of the dodder stem between the host and soil dies, and a network of threadlike stems winds through the canopy of the original host plant and may also spread to adjacent plants. Each dodder plant produces thousands of hard seed that can remain dormant in the soil for at least 5 years; some of the seed may germinate as soon as they fall to the ground if conditions are favorable.

Dodder commonly invades fields from weedy or natural areas, but seed can also be introduced in water, in contaminated crop seed, and in the manure of livestock that have eaten infested forage. Dodder is a problem in alfalfa and often appears in potatoes following an infested alfalfa crop or in fields grazed by cattle that have eaten infested alfalfa. Dodder does not always reappear in the same places each year, but once seed are present, the potential for infestation remains indefinitely. Keep a permanent record of dodder infestations and survey the area regularly every season for reappearance of dodder seedlings. Use hand-cultivation to destroy seedlings before they attach to potato vines.

In infested fields, apply herbicides to control seedlings before they emerge or as soon as they emerge. If infestations are not found until the dodder is established in the vines, destroy the infested plants. If only a few plants are infested, remove them from the field and burn them. Spray larger areas with a contact herbicide, then burn the plants after they dry out. Remove or treat plants several feet beyond the apparent edge of the infestation to be sure all dodder is destroyed. Use these control measures in fence rows, ditchbanks, and road-sides as well as in the field. Destroy dodder before it produces seed; infestations left until harvest will spread seed through the field on harvesters or other equipment. Revisit treated spots every 2 weeks to kill any survivors or new seedlings, and mark the spots so they can be checked in following seasons.

Rotations of nonhost crops, such as cereals, beans, or corn, can reduce dodder populations, but it may be necessary to keep the field in a nonhost crop for several years to have a significant effect.

K.   Threadlike, leafless, yellow dodder stems twine around host plants, creating a tangled mat.

L.   Dodder flowers and seed capsules are borne in clusters. Each flower is about ⅛ inch (3 mm) long. Destroy dodder infestations before this stage to prevent seed production.

Annual Broadleaf Weeds—
Potato Family (Solanaceae)

Several potato family weeds are common pests in potato fields. Because they are closely related to potatoes, they are hosts for several potato diseases and insect pests, including all potato viruses, *Tobacco rattle virus* (TRV), Verticillium wilt, late blight, black dot, powdery scab, nematodes, Colorado potato beetle, and green peach aphid.

Potato Volunteers, *Solanum tuberosum*

In rotation crops, field borders, or cull piles, potato volunteers are a serious problem because they can be reservoirs of potato insects, nematodes, and diseases. Volunteers are often a major source of *Potato leafroll virus* and other potato viruses and may also be important sources of ring rot or late blight. Populations of Colorado potato beetle or other harmful insects may build up on volunteers.

In winter potato areas, desiccation during summer fallow periods effectively controls volunteers. A hard winter freeze in the absence of snow cover is one of the most effective controls for volunteers. Tubers are protected against mild winters if buried 1 inch (2.5 cm) deep and against severe winters if buried 4 to 6 inches (10 to 15 cm) deep.

K. Dodder infestation

L. Dodder flowers and seed capsules

Most available herbicides do not control potato volunteers. Check references listed in the suggested reading as well as with local authorities for currently available materials that are effective. Apply translocated or contact herbicide to volunteers along field borders or to isolated volunteers in rotation crops. Foliar application of a sprout inhibitor greatly reduces volunteers in the following crop. Eliminate cull piles or treat plants that emerge from them with an effective herbicide.

Nightshades, *Solanum* spp.

Three nightshade species can be problems in potatoes. The worst of these weeds in most potato-growing areas of the West is hairy nightshade, *Solanum sarrachoides*. Black nightshade, *Solanum nigrum*, is more common in the growing areas of Arizona, central and southern California, and New Mexico. Black and hairy nightshades may be found together; the two species can be distinguished by the shape of the calyx, the color or appearance of mature berries, and the hairiness of stems and leaves. Cut-leaf nightshade, *Solanum triflorum*, occurs sporadically throughout the western states. Being close relatives of potato, nightshades host many pathogens, nematodes, and insect pests of potatoes.

The best strategy for controlling nightshades is to plan a crop rotation sequence that prevents populations from building up. Choose alternate crops such as corn, sorghum, cereals, or sugar beets that can be treated with herbicides that kill nightshades. Nightshades are prolific seed producers; once a nightshade population builds up, it may take several years' rotations to reduce infestations significantly. Hairy nightshade produces so many seed, up to 45,000 per plant, that large numbers of seedlings may survive even when an herbicide is 99% effective. Although most seed germinate after 1 or 2 years, some may survive for up to 40 years.

Combinations of two or three different herbicides or multiple applications may be necessary to control nightshade infestations in potatoes. Make sure cultivations and herbicide applications do not allow nightshades to develop beyond the seedling stage.

Sanitation is essential to prevent nightshades from building up in fields free of heavy infestations. Destroy nightshade plants before they mature, because seed from a single plant can produce thousands of seedlings in subsequent seasons. Rogue any plants that have set fruit and remove them from the field to prevent spreading the seed during harvest. Apply a translocated herbicide to nightshades growing along field borders; weeds in field borders are potential sources for field infestations as well as disease and insect problems.

M. Hairy nightshade seedling

N. Hairy nightshade fruit

O. Black nightshade seedling

M.   Seed leaves of hairy nightshade are lance shaped; the first true leaves have prominent veins, smooth or serrated margins, and numerous fine, short hairs, especially along the underside of the main vein.

N.   Hairy nightshade berries are green or yellowish brown when mature; they are occasionally brownish black. The calyx covers the entire upper surface of the fruit. The pedicels, like stems and leaves, are usually conspicuously hairy. Mature plants can be up to 2 feet (60 cm) tall, and usually form large, spreading mats.

O.   Seed leaves of black nightshade are elongate-oval and pointed; the first true leaves are spade shaped with smooth edges. Lower surfaces are often purple.

P. Black nightshade berries turn from green to black when mature, and the calyx covers only a small part of the fruit surface. Petioles, stems, and leaves have some hairs but are not densely hairy or sticky. Black nightshade plants may be erect and bushy, up to 3 feet (1 m) tall, or prostrate and spreading.

Q. Cutleaf nightshade leaves are deeply cleft at the edges. The plant may be erect or prostrate, up to 2 feet (0.6 m) tall, and smooth or somewhat hairy. The berries (not shown) are green or yellowish and contain many yellow seed.

Groundcherries, *Physalis* spp.

Groundcherries are summer weeds that occur in most areas of the western states, more commonly in the growing areas of southern California and Arizona. Groundcherries are related to nightshades and are also hosts for many potato pests. They are controlled by the same practices used for nightshades.

R. Groundcherry flowers are cup shaped and about ⅜ inch (9 mm) across. All species have a smooth, tomatolike fruit enclosed in a characteristic green, heart-shaped calyx.

Annual Broadleaf Weeds— Mustard Family (Brassicaceae)

The mustard family contains several common weeds that are generally early-season or winter annuals or biennials, and may grow all year long in cool areas. Mustard family weeds are hosts for TRV, green peach and other aphids, and flea beetles. Green peach aphid and other aphid vectors of potato viruses may build up on these weeds in the spring and then migrate into potato fields later.

Mustard family weeds generally are controlled in potato crops with routine cultivations and herbicides. Use translocated herbicides to control mustard family

P. Black nightshade fruit

Q. Cutleaf nightshade

R. Groundcherry flower and fruit

S. Mustard seedling

T. Birdsrape mustard

U. London rocket seedling

weeds in field borders and other noncrop areas, being sure to avoid spray drift.

Mustards, *Brassica* spp.

Mustards are common winter or early-spring weeds. Some species may grow as biennials. They are not usually serious pests in potatoes, but they can become a problem in winter potatoes in the low desert valleys if not controlled before vines close. Mustards are early-season hosts for buildup of aphid populations in orchards or field borders.

S. All mustard seedlings in the *Brassica* genus have broad seed leaves with a deep notch at the tip. The first true leaves are bright green on the upper surface and paler below.

T. Most mature mustards have dense clusters of yellow flowers at the tips of branches. Flowers have 4 petals. Leaves are toothed, alternate, and often deeply lobed, especially toward the base of the plant. The species shown here is birdsrape mustard, or wild turnip, *Brassica rapa*.

London Rocket, *Sisymbrium irio*

A winter annual, London rocket is more common in Arizona and California than in other potato-growing areas of the West. It is a highly competitive weed in grain rotations.

U. The edges of the first true leaves of London rocket seedlings are often somewhat indented. They are very difficult to distinguish from shepherd's-purse seedlings. Most or all of the early leaves of London rocket are deeply indented; early leaves of shepherd's-purse can be either indented or smooth edged.

V. London rocket

W. Tumble mustard

Y. Shepherd's-purse

Z. Field pennycress

V. London rocket flowers form small clusters at the tips of stems that bear long, slender seed pods. The plants usually grow to about 2 feet (0.6 m) tall.

Tumble Mustard, *Sisymbrium altissimum*

Tumble mustard, also called Jim Hill mustard, is a winter annual weed that is most common in Northwest growing areas (Idaho, Montana, Oregon, Washington, northern Nevada, and northern California). It is frequently a problem in grain rotations. In the Klamath Basin, tumble mustard is the earliest green peach aphid host after peach trees.

W. Tumble mustard plants grow 2 to 5 feet (0.6 to 1.5 m) tall. They branch widely, and the upper leaves have a thin, fingerlike appearance. Tumble mustard seedlings are similar to London rocket, but early leaves have conspicuous, coarse hairs.

Shepherd's-purse, *Capsella bursa-pastoris*

Shepherd's-purse is a common weed throughout the western states.

X. The margins of the first true leaves of shepherd's-purse seedlings are not indented and are covered with hairs that have 4 or 5 rays arranged in a starlike pattern when viewed with a hand lens.

Y. Shepherd's-purse grows 3 to 18 inches (7.5 to 45 cm) tall from the center of a rosette of indented and smooth-edged leaves. The heart-shaped seed pods make this species easy to recognize when it is mature.

X. Shepherd's-purse seedling

Field Pennycress, *Thlaspi arvense*

A common weed in Northwest growing areas, field pennycress may be a serious pest in grain rotations.

Z. Field pennycress has clusters of small white flowers that branch out from the tips of short plants, about 6 to 15 inches (15 to 38 cm) tall. The seed pods have a flat, palm fan shape with a notch in the tip.

AA. Flixweed seedling

BB. Tansymustard

CC. Lambsquarters seedling

DD. Lambsquarters

Flixweed, Tansymustard *Descuraina* spp.

Flixweed, *Descuraina sophia*, and tansymustard, *D. pinnata*, are winter annuals that occur throughout the western states. They are usually more abundant in the drier growing areas.

AA. Seedlings of flixweed (shown here) and tansymustard are very similar. Seed leaves are linear. The first true leaves are deeply lobed, and subsequent leaves have a fernlike appearance.

BB. Flixweed and tansymustard grow up to 2 feet (0.6 m) tall, with yellow flowers at the ends of long, slender stems. These stems bear seed pods that are about ½ inch (12 mm) long in tansymustard (shown here) and ½ to 1 ½ inch (1.3 to 4 cm) long in flixweed. The seed pods of tansymustard are shorter than their petioles. The finely divided leaves of flixweed and tansymustard distinguish them from other mustard family weeds.

Annual Broadleaf Weeds— Goosefoot Family (Chenopodiaceae)

Weeds of the goosefoot family are common summer annuals in all potato-growing areas of the West. Cultivations and soil-applied herbicides generally control these weeds.

Lambsquarters, *Chenopodium album*

A common weed throughout the western states, lambsquarters is sometimes called white goosefoot or fat-hen. Lambsquarters is a winter weed in the low desert valleys of California and Arizona, where it can be a problem if not controlled before vines close; elsewhere it is a summer weed. Lambsquarters is a host for the beet leafhopper, green peach aphid, and *Potato virus* X (PVX).

CC. Lambsquarters seed leaves are narrow, with nearly parallel sides. The seed leaves and early true leaves are dull bluish green above and often purple below. Newly emerging leaves sometimes have a pink color in the center. The mealy leaf texture distinguishes lambsquarters from most other seedlings.

DD. Lambsquarters may grow up to 6 feet (1.8 m) tall, depending on moisture and soil fertility. Plants are erect and highly branched, with stems often streaked with pink or purple. Tiny, inconspicuous flowers are packed in dense clusters at the tips of the main stem and branches.

EE. Kochia seedling

FF. Kochia

Kochia, *Kochia scoparia*

Kochia is a common summer weed in the Rocky Mountain, Snake River Valley, Columbia Basin, and Klamath Basin growing areas. Populations of kochia that are resistant to ALS inhibitor herbicides (e.g., sulfonylureas and imidazolinones) occur commonly.

EE. Kochia seed leaves (not visible) are thick, elongated, and often magenta underneath. The first true leaves are grayish, hairy, and borne in tight rosettes.

FF. Mature kochia plants are 20 to 60 inches (0.5 to 1.5 m) tall, with a bushy appearance. Small, inconspicuous flowers are formed at the ends of branches. Stems of kochia may turn reddish in the fall.

Russian Thistle, *Salsola tragus*

Russian thistle occurs throughout the western states, usually more abundantly in drier areas. It can be a serious pest of potatoes in the Columbia Basin and Rocky Mountain growing areas. Russian thistle is the preferred host of the beet leafhopper.

GG. The seed leaves and first true leaves of Russian thistle are long and slender and resemble pine needles.

HH. Mature Russian thistle plants are spherical bushes up to 5 feet (1.5 m) tall, with small, lance-shaped leaves. Stems usually are striped with red or purple. When old, the plants turn grayish brown, break off at the soil line, and become "tumbleweeds," spreading their seed as they are blown about in the wind.

GG. Russian thistle seedling

HH. Russian thistle

II. Annual sowthistle
seedling

JJ. Annual sowthistle

Annual Broadleaf Weeds— Sunflower Family (Asteraceae)

Weeds in the sunflower or aster family may be problems in some potato-growing areas. The perennial Canada thistle is important in many areas. Some sunflower family weeds can host certain potato viruses. The weed species in this family vary in their susceptibilities to herbicides used in potatoes and may be difficult to control with chemicals.

Annual Sowthistle, *Sonchus oleraceus*

Annual sowthistle is most common in the coastal valleys of Oregon and Washington, the Sacramento–San Joaquin Delta, and in the low desert valley growing areas of Arizona and California. It is a winter or early spring weed in most potato-growing areas. It grows year-round in the Delta and San Joaquin Valley growing areas of California. Control sowthistle with early-season cultivations or herbicide applications before the weed emerges.

II. Seed leaves of annual sowthistle have smooth edges and a grayish powder on the upper surface. The first true leaves are broad at the tip and taper at the base. They have slightly wavy edges and may have soft prickles around their margins.

JJ. Mature annual sowthistle plants are 1 to 4 feet (0.3 to 1.2 m) tall. Yellow flowers are borne on the ends of stalks. The base of each long, pointed leaf wraps around the stem. If stems or leaves are cut or torn, a milky juice flows out.

Common Sunflower, *Helianthus annuus*

Common sunflower is a summer weed that occurs sporadically in Rocky Mountain, Snake River Valley, Columbia Basin, and low desert valley growing areas. It is a serious pest

in Colorado and is a common problem in grain rotations wherever it occurs. Sunflower is a host for PVX and TRV. Try to control sunflowers with early-season cultivations and preemergence herbicides; they are difficult to control later in the season. Mature sunflower plants grow from 1.5 to 9 feet (0.5 to 2.7 m) tall and are highly branched. The characteristic yellow flowers make sunflower easy to recognize.

KK. Seed leaves of common sunflower are twice as long as they are wide. The first true leaves are similar in shape, with smooth edges and short, rough hairs. Sunflower seedlings have strong taproots.

Common Cocklebur, *Xanthium strumarium*

Although it is a summer weed that occurs throughout the western states, common cocklebur is generally not a serious pest of potatoes. However, when present in potato it is difficult to control. Mature cocklebur plants are bushy, from 2 to 5 feet (0.5 to 1.5 m) tall, with heart-shaped leaves and dark red spots or streaks on the stems. The leaves have a rough texture. Female flowers develop into football-shaped burs with stiff, hooked spines. Each bur contains two black seed. Cocklebur is a host for TRV.

KK. Common sunflower seedling

LL. Cocklebur seedling

LL. The seed leaves of cocklebur are narrow, long, and bright green. The first true leaves are much broader, with notched edges and a dull green upper surface. Cocklebur seed and seedlings are toxic to livestock.

Other Annual Broadleaf Weeds

Pigweeds, *Amaranthus* spp.

The pigweeds are dominant summer weeds in most potato-growing areas. Redroot pigweed, *Amaranthus retroflexus*, is the most common and is one of the most prevalent weeds in all potato-growing areas. Redroot pigweed, also called rough pigweed, is a host for the beet leafhopper and for PVX. Other pigweed species that may occur in potatoes include Powell amaranth (*A. powellii*), tumble pigweed (*A. albus*), prostrate pigweed (*A. blitoides*), and smooth pigweed (*A. hybridus*), which is also called green amaranth. Pigweeds are controlled with cultivations and herbicides commonly used in potatoes; prostrate pigweed is the most difficult to control.

MM. Seed leaves of redroot pigweed are long, narrow, and bright magenta below. The first true leaves are oval with a shallow notch at the tip. Seedlings of other pigweed species are similar.

NN. Redroot pigweed may be 7 feet (2.1 m) tall in some situations. Branches rise mainly from the base; stems are furrowed. Leaves may be several inches long, with long petioles and prominent veins on the lower surface. The small, green flowers are arranged in dense spikes at the tops of the main stem and branches, and in smaller clusters in leaf axils.

Mallow, *Malva* spp.

Little mallow, *Malva parviflora*, occurs in the potato-growing areas of central California and the low desert valleys. It can grow year-round in these areas. Common mallow, *M. neglecta*, occurs in other areas, where it is a summer annual or biennial weed. Both species are often called cheeseweed, and they are difficult to distinguish from one another. Cultivate to destroy mallow seedlings while they are small; otherwise they will rapidly form deep, tough taproots. Mallow is difficult to control with available herbicides.

OO. Seedlings of little mallow (shown here) and common mallow are similar. Seed leaves are triangular or heart shaped. True leaves are round or kidney shaped, with scalloped edges.

PP. Common mallow can grow up to 4 feet (1.2 m) tall; little mallow (shown here) can be prostrate or grow up to 5 feet (1.5 m) tall. These weeds are frequently called cheeseweed because the distinctive fruits resemble tiny wheels of cheese.

MM. Redroot pigweed seedling

NN. Redroot pigweed

OO. Little mallow seedling

PP. Little mallow flowers and fruit

JOSEPH M. DI TOMASO

QQ. Burning nettle seedlings

RR. Burning nettle

Nettle, *Urtica urens*

Burning nettle, also called stinging nettle, is a problem in the San Joaquin Valley of California where it is a winter weed. Encourage nettle seedlings to emerge before potatoes by applying a light irrigation if there is no rain, then remove them with cultivation. Use a contact herbicide if cultivations are not possible.

QQ.   Seed leaves of burning nettle are round or slightly elongated with smooth edges and a small notch in the tip. The first true leaves have distinctly toothed margins.

RR.   Mature nettle plants are 5 to 24 inches (12 to 60 cm) tall, with stems branching at the base. Stems

are square in cross-section; both leaves and stems have stinging hairs.

Wild Buckwheat, Common Knotweed *Polygonum* spp.

Wild buckwheat, *Polygonum convolvulus*, and common knotweed, *P. arenastrum*, also called prostrate knotweed, are common weeds in many potato-growing areas. They are controlled with cultivation and available herbicides.

SS.   True leaves of wild buckwheat are shaped like arrowheads and are more pointed than leaves of field bindweed, with which it is sometimes confused. Mature plants have trailing or twining stems 8 to 40 inches (20 to 100 cm) long. The inconspicuous flowers form clusters at the base of leaf stalks or at the end of stems.

TT.   Seed leaves of common knotweed are long, narrow, rounded at the tip, and have whitish streaks or blotches. The true leaves are much broader and emerge from a membranous sheath that encircles the stem. Nodes are usually swollen.

SS. Wild buckwheat

TT. Common knotweed seedling

UU. Mature knotweed plants may be prostrate or erect. Stems are thin, tough, and extensively branched. The tough stems may become entangled in cultivation equipment.

Common Purslane, *Portulaca oleracea*

Common purslane occurs in all potato-growing areas and is usually controlled with cultivation and available herbicides. Plants can reroot after cultivation if they remain in contact with moist soil.

VV. Seed leaves of common purslane are smooth and succulent with a reddish tinge. The first true leaves have rounded tips that are broader than their bases. Newly emerging leaves on young plants are flattened together.

WW. Mature purslane plants form mats of highly branched, reddish stems that are 6 inches to 3 feet (15 to 90 cm) long and turned up at their tips. The pale yellow, cup-shaped flowers are usually open just in the morning.

Annual Grasses (Poaceae)

Many annual grasses, including volunteer grains, are common weeds in potato fields. Most are controlled with available herbicides; repeated applications may be necessary where grasses are prevalent, if shading by potato vines is not sufficient to prevent the weeds from developing and setting seed. Corn, wheat, and barley volunteers are most difficult to control. Grazing livestock on stubble after a corn harvest reduces the number of corn volunteers.

Barnyardgrass, *Echinochloa crus-galli*

Also called watergrass, barnyardgrass is one of the most common summer annual grass pests. It occurs in all potato-growing areas.

XX. The best way to recognize barnyardgrass is to strip off leaves and look at the collar region. It is the only common summer weedy grass that lacks a ligule.

VV. Common purslane seedling

WW. Common purslane

UU. Common knotweed

XX. Barnyardgrass collar

YY. Barnyardgrass

YY. Barnyardgrass plants can grow up to 6 feet (1.8 m) tall where water is available in the summer, but may be only 6 inches (15 cm) tall where it is drier. Some plants may root at the lower nodes, forming large clumps. Flower heads of some varieties have long bristles. Each barnyardgrass plant can produce a large quantity of seed.

Foxtails, *Setaria* spp.

Summer weeds throughout the western states, foxtails are more common in the northern growing areas. Yellow foxtail, *Setaria pumila*, and green foxtail, S. *viridis*, are distinguished from each other and from other summer grasses by characteristics of the collar and leaf sheath. Foxtails, sometimes called bristlegrass or pigeon grass, are frequent pests of alfalfa.

ZZ. Yellow foxtail

ZZ. Mature foxtail plants are 1 to 3 feet (0.3 to 0.9 m) tall, with branching and some spreading at their bases. Leaf blades are 4 to 15 inches (10 to 38 cm) long, and most have a spiral twist. Long hairs are present at the base of the leaf blade. Flower heads of yellow foxtail (shown here) are dense spikes with yellow or reddish bristles that are ¼ to ⅓ inch (6 to 8.5 mm) long. Green foxtail flower spikes have green or purple bristles.

AAA. The ligule of yellow foxtail is a fringe of hairs, and there are no auricles. There are no hairs on the leaf sheath margin below the collar as on green foxtail.

BBB. If you pull the leaf sheath of green foxtail away from the stem, you can see fine hairs on the leaf

AAA. Yellow foxtail collar

BBB. Green foxtail collar

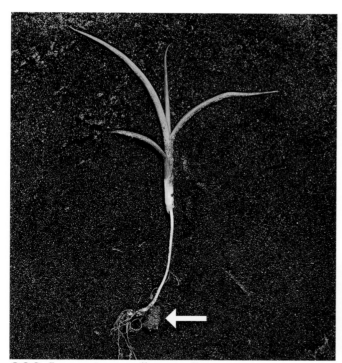

CCC. Longspine sandbur seedling

DDD. Longspine sandbur

EEE. Wild oat

sheath below the collar region. The growth habit of green foxtail is similar to yellow foxtail, but the leaf blades of green foxtail are flat and lack hairs.

Longspine Sandbur, *Cenchrus longispinus*

Sandbur occurs sporadically throughout the western states, growing in sandy soils. It may be a problem in the Columbia Basin, Colorado, New Mexico, and Wyoming.

CCC. Sandbur seedlings closely resemble those of barnyardgrass. However, if you carefully dig up a sandbur seedling, you can often find the bur still attached (arrow).

DDD. Sandbur stems are weak and branched, and the plant often spreads along the ground. Sandbur plants sometimes have a dense, bushy growth. Flower heads consist of several burs (as many as 40) in a spike that is partially enclosed by a leaf sheath. The burs are yellowish green when young and turn a light brown when they mature.

Wild Oat, *Avena fatua*

A pest throughout the western states, wild oat is most serious in the growing areas of northern California, Oregon, Washington, and Idaho. Wild oat seed are a favorite forage of meadow mice; where meadow mice become a problem in potatoes, their burrows are frequently found near wild oat plants. Wild oat can be distinguished from cultivated oat by the twisted awns that are bent at right angles when flowers are mature and by the presence of a horseshoe-shaped scar ("muleshoe") at the base of the seed.

EEE. Wild oat plants are 1 to 4 feet (0.3 to 1.2 m) tall, with several stems growing from the base. The flowers are large and open, resembling those of cultivated oats.

FFF.   A large, whitish ligule is present on the collar of a wild oat leaf blade. Few or no hairs are present on the leaf sheath; hairs are generally present on the margins of the leaf blade.

Witchgrass, *Panicum capillare*

A summer annual weed, witchgrass is sometimes a problem in potato-growing areas of Colorado, the Columbia Basin, and the Snake River Valley.

GGG.   The mature witchgrass plant is bushy, branched from the base, and has a fuzzy appearance.

HHH.   The stem, sheath, and leaf blade of witchgrass are covered with long, coarse hairs. The ligule consists of a fringe of hairs.

GGG. Witchgrass

FFF. Wild oat collar

HHH. Witchgrass collar

Voles, *Microtus* spp., are 4 to 6 inches (10 to 15 cm) long when full grown, with a blunt nose, small eyes, and short, furry ears. The tail is less than half as long as the body and is slightly hairy.

Vertebrates

Vertebrates generally are not significant pests of potatoes in the western United States. In many growing areas, minor tuber damage may be caused every year by pheasants, geese, ground squirrels, or voles, but control measures are not necessary. In some locations, particularly in the Northwest coast areas, pocket gophers may be a problem if they have built up in a rotation crop; controls are generally not applied in potatoes. Controls are required for voles in parts of the Klamath Basin during the occasional years when mouse populations reach damaging levels.

Voles, *Microtus* spp., also called meadow mice or field mice, live in colonies in areas such as irrigated pastures, fencerows, or weedy ditchbanks, where the soil is suitable for burrowing and where vegetation provides cover. They usually avoid the sandy soils in which potatoes are commonly grown in the western states. The soil of the Tule Lake Basin of northern California is more favorable for voles, and there is also abundant natural habitat where vole populations can remain active year-round. In this area vole populations reach levels that require control every 7 to 10 years. Controls are required less frequently in other parts of the Klamath Basin.

Voles move into potato fields when infested grain or alfalfa fields are harvested, usually in August or September, burrowing into hills for shelter and to feed on tubers. Their feeding damages tubers directly and their burrows may expose tubers to sunlight or freezing temperatures, causing additional losses. Damage may also be caused by predators that dig into potato hills in search of voles. If vole populations are high at harvest, voles will be carried into storage, where they will continue to feed on tubers.

In the Tule Lake Basin, ditchbanks are monitored for voles each spring by the county agricultural commissioners' offices. When trap monitoring or visual estimates of vole activity indicate damage is likely, poison baits may be applied to ditchbanks in the spring as a preventive measure, or directly to potato fields after vine death. If monitoring programs in your area indicate that voles may become serious, begin checking your fields for vole activity when removing irrigation pipe at the end of the season or after nearby grain or alfalfa fields are harvested. Look for active burrows—ones with fresh soil around the entrance. Check closely around clumps of wild oats, where vole burrows are most likely to be found.

In other Klamath Basin growing areas, there are no regular monitoring programs because voles rarely reach damaging levels. However, a routine check for vole activity near the end of each season is a good way to identify the occasional year when voles may become a problem. If you find vole activity in your fields at the end of the season, check with your local county agent, farm advisor, or agricultural commissioner's office to see if bait application is recommended.

Voles dig small, shallow burrows. These burrows will be found anywhere voles are present. In potato fields, voles burrow to feed on tubers. Active burrows have fresh soil around the entrance.

Voles damage potatoes by feeding directly on tubers. Their burrowing activity also may expose tubers to sunlight or freezing temperatures.

Suggested Reading

General

Best Management Practices for Potato Production in the San Luis Valley. 1996. Publication XCM-196 (Colorado).

Characteristics of Potato Varieties in the Pacific Northwest. 1993. Publication PNW 454 (Idaho, Oregon, Washington).

Commercial Potato Production in North America, Potato Association of America Handbook. 1994. Potato Association of America, Orono, ME.

Cull and Waste Potato Management. 2001. Current Information Series No. 814 (Idaho).

Cultural Management of Bannock Russet Potatoes. 2002. Current Information Series No. 1103 (Idaho).

Cultural Management of Gem Russet Potatoes. 2001. Current Information Series No. 1103 (Idaho).

Cultural Management of Ranger Russet Potatoes. 1998. Current Information Series No. 919 (Idaho).

Cultural Management of Russet Norkotah Potatoes. 2003. Current Information Series No. 1106 (Idaho).

Potato Association of America. On the World Wide Web at www.umaine.edu/paa.

Potato Growers Newsletter. Potato Lab, P.O. Box 172060, Montana State University, Bozeman, MT 59717-2060.

Potato Health Management. 1993. R. C. Rowe, ed. American Phytopathological Society, St. Paul, MN.

Potato Information Exchange. On the World Wide Web at http://oregonstate.edu/potatoes.

Potato Production Systems. 2003. J. C. Stark and S. L. Love, eds. (Idaho).

Potato Progress. Washington State Potato Commission. On the World Wide Web at www.potatoes.com/research/potatoprogress/.

Potato Seed Management: Seed Size and Age. 1995. Current Information Series 1031 (Idaho).

Potatoes: Influencing Seed Tuber Behavior. 1992. Publication PNW 248 (Idaho, Oregon, Washington).

Proceedings of the University of Idaho Winter Crop Schools. Published annually (Idaho).

San Luis Valley Potato Production Manual. Publication XCM-181 (Colorado).

Soil, Plant and Water Analytical Laboratories for Montana Agriculture. 2005. Publication 4481 (Montana).

Specialty Potato Production and Marketing in Southern Idaho. 2003. Current Information Series No. 1110 (Idaho).

Specific Gravity of Potatoes. 1984. Current Information Series No. 609 (Idaho).

Using Green Manures in Potato Cropping Systems. 2003 Extension Bulletin 1951E (Washington).

Washington State Potato Conference and Trade Show. Proceedings published annually by the Washington State Potato Commission, 108 Interlake Road, Moses Lake, WA 98837.

Integrated Pest Management

Beneficial Organisms Associated with Pacific Northwest Crops. 1993. Publication PNW 343 (Idaho, Oregon, Washington).

Best Management Practices for Integrated Pest Management in the San Luis Valley. 1996. Publication XCM-195 (Colorado).

High Plains Pest Management Guide. On the World Wide Web at http://highplainsipm.org/.

Integrated Crop and Pest Management Record-Keeping System Record Book. 1988. Extension Bulletin 12 (Montana).

IPM in Practice: Principles and Methods of Integrated Pest Management. 2001. Publication 3418 (California).

Treasure Valley and Pacific Northwest Pest Alert Network. On the World Wide Web at http://www.tvpestalert.net/.

UC IPM Pest Management Guidelines: Potato. Revised continuously. Publication 3463 (California). Also available from University of California Cooperative Extension offices and at the UC IPM World Wide Web site, http://www.ipm.ucdavis.edu.

Soil and Water

AgriMet: Pacific Northwest Cooperative Agricultural Weather Network. On the World Wide Web at http://www.usbr.gov/pn/agrimet/.

Best Management Practices for Irrigation Management. 1994. Publication XCM-173 (Colorado).

CIMIS: California Irrigation Management Information System. On the World Wide Web at http://www.cimis.water.ca.gov/.

Diagnosing Soil Physical Problems. 1976. Publication 2664 (California).

Irrigation Scheduling. 1994. Publication PNW 288 (Idaho, Oregon, Washington).

Measuring Irrigation Water. 1959. Publication 2956 (California).

Oregon Crop Water Use and Irrigation Requirements. 1999. Publication EM 8530 (Oregon).

Potato Irrigation Management. 1997. Bulletin 789 (Idaho).

Potato Production with Limited Water Supply. 2004. Current Information Series No. 1122 (Idaho).

Principles of Soil Sampling for Northwest Agriculture. 1981. Western Regional Extension Publication 9. (Idaho, Oregon, Washington).

Soil Sampling. 1998. Bulletin 740 (Idaho).

Soil Sampling. 1999. Publication MT 8602 (Montana).

Soil Test Interpretations. 2000. Guide A-122 (New Mexico).

Soil Water Monitoring and Measurement. 1999. Publication PNW 475 (Idaho, Oregon, Washington).

Nutrients

Best Management Practices for Nitrogen Fertilization. 1994. Publication XCM-172 (Colorado).

Best Management Practices for Phosphorous Fertilization. 1994. Publication XCM-175 (Colorado).

Critical Nutrient Ranges in Northwest Crops. 1980. Western Regional Extension Publication 43. (Idaho, Oregon, Washington).

Fertilization and Liming. On the World Wide Web at http://oregonstate.edu/potatoes/fertilizer.htm.

Fertilizer Guide for New Mexico. 1990. Publication A-128 (New Mexico).

Fertilizer Guidelines for Montana Crops. 2005. Extension Bulletin 161 (Montana).

Fertilizer Placement. 2001. Current Information Series No. 757 (Idaho).

Fertilizing Potato. 1992. Publication SF-715, North Dakota Extension Service.

Fertilizing with Manure. 2004. Publication PNW 533 (Idaho, Oregon, Washington).

Interpretation of Soil Test Nitrogen: Irrigated Crops in Central Washington. 1979. Publication EM 3076 (Washington).

Monitoring Soil Nutrients Using a Management Unit Approach. 2003. Publication PNW-570-E (Idaho, Oregon, Washington).

Nitrogen Uptake and Utilization by Pacific Northwest Crops. 1999. Publication PNW 513 (Idaho, Oregon, Washington).

Nutrient Management Guide: Central Washington Irrigated Potatoes. 1999. Extension Bulletin 1882 (Washington).

Nutrient Management Guidelines for Russet Burbank Potatoes. 2004. Bulletin 840 (Idaho).

Phosphorus Fertilization of Irrigated Soils in Central Washington. 1970. Extension Bulletin 602 (Washington).

Phosphorus Fertilizer: Broadcast, Banding, and Starter. 1991. Extension Bulletin 1637 (Washington).

Potato Fertilization on Irrigated Soils. 1991. Publication 03425 (Minnesota). On the World Wide Web at http://www.extension.umn.edu/distribution/cropsystems/DC3425.html.

Potato Nutrient Management for Central Washington. 1999. Extension Bulletin 1871 (Washington).

Sampling for Plant Tissue Analysis. 1999. Guide A-123 (New Mexico).

Soil and Plant Tissue Testing in California. 1983. Publication 1879 (California).

Soil Sampling as a Basis for Fertilizer Application. 1998. Publication SF-990 (North Dakota).

Soil Test Interpretation Guide. 1999. Extension Circular 1478 (Oregon).

Tissue Analysis—A Guide to Nitrogen Fertilization for Russet Burbank Potatoes. 1984. Current Information Series 743 (Idaho).

Seed Certification

California Certified Seed Potatoes. Article 7, California Department of Food and Agriculture, 1220 N Street, Sacramento, CA 95814.

Colorado Rules and Regulations for Certified Seed Potatoes. Potato Certification Service, Department of Horticulture, Colorado State University, Fort Collins, CO 80523.

Idaho Rules of Certification. Idaho Crop Improvement Association, Inc., 1680 Foote Drive, Idaho Falls, ID 83402. Also on the World Wide Web at www.idahocrop.com/.

Montana State University Seed Potato Certification Rules and Regulations. Potato Lab, P.O. Box 172060, Montana Stare University, Bozeman, MT 59717-2060.

Oregon Certified Seed Handbook. (Oregon).

Oregon Potato Seed Certification Standards. (Oregon).

Potato Seed Management: Seed Certification and Selection. 2000. Current Information Series 974 (Idaho).

Seed Potatoes: Certification Requirements and Standards. Utah Crop Improvement Association, Utah State University, UMC 48, Logan, UT 82322.

Harvest and Storage

Commercial Application of CIPC Sprout Inhibitor to Storage Potatoes. 1997. Current Information Series No. 1059 (Idaho).

Insulation and Vapor Barriers in Potato Storage Buildlings. 1987. Publication PNW 236 (Idaho, Oregon, Washington).

Organic and Alternative Methods for Potato Sprout Control in Storage. 2004. Current Information Series No. 1120 (Idaho).

Potato Bruise Prevention: Cleaning and Disinfecting Potato Storages. 1999. Video #876 (Idaho).

Potato Bruise Prevention: Handling. 1990. Video #586, available in English or Spanish (Idaho).

Potato Bruise Prevention: Harvester Chain Adjustment. 1989. Video #471, available in English or Spanish (Idaho).

Potato Bruise Prevention: The Harvester. 1985. Video #275, available in English or Spanish (Idaho).

Potato Bruise Prevention: Reducing Bruising in Fresh-Pack Warehouses. 1996. Video #794 (Idaho).

Potato Storage and Ventilation. 1993. Publication EM 2799 (Washington).

Potatoes: Storage and Quality Maintenance in the Pacific Northwest. 1985. Publication PNW 257 (Idaho, Oregon, Washington).

Reducing Potato Damage During Harvest. 1981. Extension Bulletin 0646 (Washington).

Reducing Potato Harvesting Bruise. 1983. Extension Bulletin 1080 (Washington).

Storage Management for Gem Russet. 2004. Current Information Series No. 1118 (Idaho).

Storage Management for Summit Russet. 2004. Current Information Series No. 1123 (Idaho).

Storage Management for Umatlla Russet. 2003. Current Information Series No. 839 (Idaho).

Vine Kill and Long-term Storage of Ranger Russet Potatoes. 2004. Current Information Series No. 1119 (Idaho).

Pesticide Application and Safety

Best Management Practices for Agricultural Pesticide Use. 1995. Publication XCM-177 (Colorado).

Calibrating Multiple Nozzle Boom Type Sprayers. 1998. Publication MP-93.2 (Wyoming).

Farm Safety Series. 1998. Publication PNW 512, available in English or Spanish (Idaho).

First Aid for Pesticide Poisoning. 1993. Publication PNW 278 (Idaho, Oregon, Washington).

National Pesticide Safety Education Core Manual. 2005. C. Randall et al., eds. U.S. Environmental Protection Agency, Office of Pesticide Programs, Washington, D.C..

Oregon Pesticide Safety Education Manual: A Guide to the Safe Use and Handling of Pesticides. 2004. Publication EM 8850 (Oregon).

Pesticide Safety: Worker Protection II. 1997. Video #846 (Idaho).

Pesticide Safety Information Series. California Department of Food and Agriculture, 1220 N Street, Sacramento, CA 95814.

Pesticide Storage and Handling. 1998. (Arizona).

Private Pesticide Applicator Training and Certification in Montana. 2000. Publication MT200012AG (Montana).

Safe and Effective Use of Pesticides. 2nd ed. 2000. Publication 3324 (California).

Safe Handling of Pesticides-Mixing. 2002. Publication MT200109AG (Montana).

Soil Fumigation. 1998. Publication MISC 163 (Washington).

Tank-Mixing Herbicides. 1984. Publication PNW 255 (Idaho, Oregon, Washington).

Insects

Colorado Potato Beetle: Organic Control Options. 2003. National Sustainable Agriculture Information Service. Available on the World Wide Web at http://attra.ncat. org/attra-pub/.

Hotline for Aphids on Potatoes. On the World Wide Web at http://www.wsu.edu/~potatoes/

Insect Pests of Farm, Garden, and Orchard. 7th ed. 1979. R. H. Davidson and W. F. Lyon. Wiley, New York.

Introduction to the Study of Insects. 6th ed. 1989. D. J. Borror, C. A. Triplehorn, and N. F. Johnson. Saunders College Publishing, Philadelphia.

Keys to Damaging Stages of Insects Commonly Attacking Field Crops in the Pacific Northwest. 1998 MS 109 (Idaho).

National Audubon Society Field Guide to North American Insects and Spiders. 1995. L. Milne and M. Milne. A. A. Knopf, New York.

Natural Enemies Handbook: The Illustrated Guide to Biological Pest Control. 1998. Publication 3386 (California).

Pacific Northwest Insect Management Handbook. Revised and published annually by the extension publications offices of Oregon, Washington, and Idaho. Also available on the World Wide Web at insects.ippc.orst.edu/pnw/insects.

Potato Tuberworm: A Threat for Idaho Potatoes. 2005. Current Information Series No. 1125 (Idaho).

Diseases and Disorders

Blackleg–Soft Rot Disease Complex in Potatoes. 1990. Current Information Series 669 (Idaho).

Brown Center and Hollow Heart in Potatoes. 1989. Extension publication 691 (Idaho).

Compendium of Potato Diseases. 2nd ed. 2001. W. R. Stevenson, ed. American Phytopathological Society, St. Paul, MN.

Control of Early Blight Disease of Potato. 1992. Current Information Series 975 (Idaho).

Corky Ringspot of Potatoes. 1992. Current Information Series 914 (Idaho).

Fungicide Application for Management of Potato Late Blight. 2001. Extension Bulletin 1923 (Washington).

Herbicide Drift and Carryover Injury in Potatoes. 1997. Publication PNW 498 (Idaho, Oregon, Washington).

Management Practices and Sugar End Potatoes. 1993. Publication PNW 427 (Idaho, Oregon, Washington).

Managing Late Blight on Irrigated Potatoes in the Pacific Northwest. 2003. Publication PNW 555 (Idaho, Oregon, Washington).

Pacific Northwest Plant Disease Management Handbook. Revised and published annually by the extension publications offices of Oregon, Washington, and Idaho. Also available on the World Wide Web at http://plant-disease.ippc.orst.edu/index.cfm.

Plant Pathology. 5th ed. 2005. G. N. Agrios. Elsevier Academic Press, Burlington, MA.

Potato Early Blight. 1994. Publication B-997 (Wyoming).

Potato Late Blight. 1996. Publication B-1032 (Wyoming).

Powdery Scab of Potato. 1995. Publication PNW 387 (Idaho, Oregon, Washington).

Silver Scurf of Potatoes. 1997. Current Information Series 1060 (Idaho).

Sugar Development in Potatoes. 1981. Extension Bulletin 0717 (Washington).

Verticillium Wilt. 2001. Extension Bulletin 1908 (Washington).

Verticillium Wilt of Potato. 1993. Current Information Series 977 (Idaho).

White Mold and Potatoes. 2003. Current Information Series 1105 (Idaho).

Nematodes

Columbia Root-Knot Nematode Control in Potato Using Crop Rotations and Cover Crops. 1999. Publication EM 8740 (Oregon).

General Recommendations for Nematode Sampling. 1981. Publication 21234 (California).

Potato Nematodes and Their Control. 1992. Current Information Series 925 (Idaho).

Potato Rot Nematode. 1990. Current Information Series 868 (Idaho).

Root-Knot Nematodes of the Pacific Northwest. 1981. PNW0190 (Washington).

Sampling for Nematodes in Soil. 1988. Extension Bulletin 1379 (Washington).

Weeds

Application of Herbicides Through Irrigation Systems. 1984. Publication MISC 0091 (Washington).

Applied Weed Science. 1985. M. A. Ross and C. A. Lembi. Burgess Publishing Co., Minneapolis.

Colorado Weed Management Guide. Revised and published annually. Publication XCM-205 (Colorado).

Commercial Horticulture Weed Control Handbook. 2003. Extension Bulletin 133 (Montana).

Cultural and Chemical Practices for Commercial Potato Weed Control. 1989. Extension Publication 695 (Idaho).

Eptam for Weed Control in Potatoes. 1998. Current Information Series 1009 (Idaho).

Flame Weeding for Vegetable Crops. 2002. National Sustainable Agriculture Information Service. Available on the World Wide Web at http://attra.ncat.org/attra-pub/.

Herbicide-Resistant Weeds and Their Management. 1993. Publication PNW 437 (Idaho, Oregon, Washington).

How to Identify Plants. 1957. H. D. Harrington and L. W. Durrell. Sage Press, Denver, 203 pp.

Metribuzin for Weed Control in Potatoes. 1991. Current Information Series No. 291 (Idaho).

Montana, Utah, Wyoming Weed Management Handbook. Revised and published periodically. Publication B-442R (Wyoming).

Montana Weed Management Handbook. Revised and published annually. Extension Bulletin 0159 (Montana).

Montana Weed Seedling Guide. 2003. Extension Bulletin 0023 (Montana).

Pacific Northwest Weed Management Handbook. Revised and published annually by the Extension publications offices of Oregon, Washington, and Idaho. Also available on the World Wide Web at http://pnwpest.org/pnw/weeds.

Selective Chemical Weed Control. 1987. Publication 1919 (California).

Volunteer Potato Control. 1998. Current Information Series 1048 (Idaho).

Volunteer Potato Management in Pacific Northwest Rotational Crops. 2005. Extension Bulletin 1993 (Washington).

Weed Management. 2003. P. J. S. Hutchinson and C. V. Eberlein. In *Potato Production Systems,* J. C. Stark and S. L. Love, eds. University of Idaho Extension.

Weed Science. 3rd ed. 1996. W. P. Anderson. West Publishing, Minneapolis-St. Paul.

Weeds of California and Other Western States. In press. Publication 3488 (California).

Weeds of the West. 2001. 9th ed. T. D. Whitson, ed. Western Society of Weed Science, Newark, CA. Also available as Publication 3350 (California).

Publications of each state are available from that state's local extension offices or from the following addresses. Western Regional publications are available from these addresses in each state.

Arizona: CAL Smart, 4042 N. Campbell Ave., Tucson, AZ 85719-1111 (http://cals.arizona.edu/pubs).

California: University of California, Agriculture and Natural Resources, Communication Services, 6701 San Pablo Avenue, 2nd Floor, Oakland, CA 94608-1239 (http://anrcatalog.ucdavis.edu).

Colorado: Cooperative Extension Resource Center, 115 General Services Building, Colorado State University, Fort Collins, CO 80523-4061 (www.cerc.colostate.edu).

Idaho: Publications, University of Idaho, P.O. Box 442240, Moscow, ID 83844-2240 (http://info.ag.uidaho.edu).

Montana: Extension Publications, Montana State University, P.O. Box 172040, Bozeman, MT 59717 (www.montana.edu/wwwpub/expubs.html).

Minnesota: On the World Wide Web at http://www.extension.umn.edu/index.html.

Nevada: College of Agriculture Publications, University of Nevada Reno, Reno, NV (www.unce.unr.edu/pubs.html).

New Mexico: On the World Wide Web at www.cahe.nmsu.edu/pubs/.

North Dakota: On the World Wide Web at http://www.ext.nodak.edu/extpubs/.

Oregon: Publication Orders, Extension & Station Communications, Oregon State University, 422 Kerr Administration, Corvallis, OR 97331-2119 (http://eesc.oregonstate.edu/).

Utah: On the World Wide Web at http://extension.usu.edu/cooperative/publications.

Washington: Bulletin Office, Cooperative Extension, Washington State University, P.O. Box 645912, Pullman, WA 99164-5912 (http://pubs.wsu.edu/cgi-bin/pubs/index.html).

Wyoming: Office of Communications and Technology Resource Center, P.O. Box 3313, Laramie, WY 82071-3313 (www.uwyo.edu/ces/pubs2.htm).

Sources of Chemicals for Nematode Host Tests

Fisher Scientific Co., www1.fishersci.com.

Mallinckrodt Baker, Inc., 222 Red School Lane, Phillipsburg NJ 08865 (www.jtbaker.com).

Sigma Chemical Corp., 3050 Spruce St., St. Louis, MO 63103 (www.sigma.sial.com).

Glossary

abiotic disorder. a disease caused by factors other than a pathogen; physiological disorder.

AMV. *Alfalfa mosaic virus.*

annual. a plant that normally completes its life cycle of germination, growth, reproduction, and death in a single year.

apical dominance. growth of the bud at the apex of a stem or tuber while growth of all other buds on the stem or tuber is inhibited.

apothecium (plural, apothecia). cup-shaped, spore-bearing structure produced by certain types of fungi such as *Sclerotinia.*

auricles. the earlike projections at the base of leaves of some grasses; used to identify species (see Fig. 40).

axil. the upper angle between a branch or leaf and the stem from which it grows.

axillary bud. a bud formed in an axil (see Fig. 4).

band application. the application of a material such as fertilizer or herbicide in strips, usually to the bed at planting or the side of the hill.

BCTV. *Beet curly top virus.*

biennial. a plant that completes its life cycle in 2 years and usually does not flower until the second season.

biotic disease. a disease caused by a pathogen, such as a bacterium, fungus, phytoplasma, or virus.

blight. a disease characterized by general and rapid killing of leaves, flowers, and stems.

BLTVA. beet leafhopper transmitted virescence agent.

broadcast application. the application of a material such as fertilizer or herbicide to the entire surface of a field.

calcareous soil. soil containing high levels of calcium carbonate.

canker. a dead, discolored, often sunken area (lesion) on a root, stem, or stolon.

caterpillar. the larva of a butterfly, moth, sawfly, or scorpionfly.

chlorophyll. the green pigment of plants that captures the energy from sunlight necessary for photosynthesis.

chlorosis. yellowing or bleaching of plant tissue that is normally green, usually caused by the loss of chlorophyll.

circulative virus. a virus that systemically infects its insect vector and usually is transmitted for the remainder of the vector's life; persistent virus.

collar region. in grasses, the region where the leaf blade and sheath meet; used in identifying species (see Fig. 40).

conidium (plural, conidia). an asexual fungus spore formed by fragmentation or budding at the tip of a specialized hypha.

control action guideline. a guideline used to determine whether pear control action is needed.

cotyledons. the first leaves of the embryo formed within a seed and present on seedlings immediately after germination; seed leaves (see Fig. 40).

curing. holding potato tubers under warm, humid conditions that favor wound healing.

degree-day. a unit of measurement combining temperature and time, used in calculating growth rates and monitoring early blight (see Table 14).

determinate. having a growth pattern in which stems stop growing at a certain developmental stage.

developmental threshold. the lowest or highest temperature at which growth occurs in a given species.

disease. a disturbance of a plant that interferes with its normal structure, function, or economic value and is typically manifested by signs or symptoms.

dormancy. a state of inactivity or prolonged rest; in potato, a period during which the buds of potato tubers will not sprout.

drag-off. the practice of removing soil from the tops of potato hills before sprout emergence.

economic threshold. a level of pest population or damage at which the cost of a control action equals the crop value gained from that control action.

epidermis. the outermost layer of living cells on the surface of a plant or animal.

evapotranspiration. the loss of soil moisture by the combination of soil surface evaporation and transpiration by plants.

eye. a collection of several buds on the surface of a potato tuber, one of which will sprout and form a new stem when conditions are favorable (Fig. 6).

field capacity. the moisture level in soil after saturation and runoff (Fig. 15).

frass. a mixture of feces and food fragments produced by an insect in feeding.

fumigation. treatment with a pesticide active ingredient that is in gaseous form under treatment conditions.

girdle. to kill or damage a ring of tissue around a stem or root; such damage interrupts the transport of water and nutrients.

glycoalkaloid. a bitter-tasting compound present in potato foliage and in the epidermis of potato tubers.

host. a living organism that is invaded by a parasite and from which the parasite obtains its food.

hypha (plural, hyphae). a tubular filament that is the structural unit of a fungus.

immune. incapable of being infected by a given pathogen.

indeterminate. having a growth pattern in which stems continue growing indefinitely.

infection. the entry of a pathogen into a host and establishment of the pathogen as a parasite of the host.

infestation. the presence of a large number of pest organisms in an area or field, on the surface of a host or anything that might contact a host, or in the soil.

inoculum. any part or stage of a pathogen, such as spores or virus particles, that can infect a host.

instar. an insect between successive molts; the first instar is between hatching and the first molt.

internode. the area of a stem between nodes.

invertebrate. an animal having no internal skeleton.

juvenile. in nematology, the immature form of a nematode that hatches from an egg and molts several times before becoming an adult.

larva (plural, larvae). the immature form of an insect, such as a caterpillar or maggot, that hatches from an egg, feeds, then enters a pupal stage.

latent. producing no visible symptoms (generally refers to an infection or a pathogen).

latent period. the time between when a vector acquires a pathogen and when the vector becomes able to transmit the pathogen to a new host.

leaf area index. the ratio between the total leaf surface area of a plant and the surface area of ground that is covered by the plant.

lenticels. natural openings in the surface of a tuber or stem, similar to leaf stomata, that can open and close and allow gas exchange (Fig. 6).

lesion. a localized area of diseased tissue, such as a canker or leaf spot.

ligule. in many grasses, a short membranous projection on the inner side of the leaf blade at the junction where the leaf blade and leaf sheath meet (see Fig. 40).

meristem. actively dividing cells of undifferentiated plant tissue such as the growing tips of roots and stems.

metamorphosis. a change in form during development.

microorganism. an organism of microscopic or small size.

micropropagation. generation of new, disease-free potato plants from tiny pieces of meristem tissue.

microsclerotia. very small sclerotia, such as those produced by the Verticillium wilt fungus.

minituber. a small tuber produced under greenhouse conditions on a small potato plant generated by micropropagation.

molt. in insects and other invertebrates, the shedding of skin before entering another stage of growth.

mutation. the abrupt appearance of a new, heritable characteristic as the result of a change in the genetic material of one individual cell.

mycelium (plural, mycelia). the vegetative body of a fungus, consisting of a mass of slender filaments called hyphae.

natural enemies. predators, parasites, or pathogens that are considered beneficial because they attack and kill organisms that we normally consider to be pests.

necrosis. death of tissue accompanied by dark brown discoloration, usually occurring in a well-defined part of a plant, such as the portion of a leaf between leaf veins or the xylem or phloem in a stem or tuber.

node. the slightly enlarged part of a stem where buds are formed and where leaves, branches, eyes, and flowers originate.

nonpersistent virus. a virus that is carried in the foregut of its insect vector and is lost after the vector feeds once or a few times.

nymph. the immature stage of insects such as aphids, leafhoppers, and psyllids that gradually acquires adult form through a series of molts without passing through a pupal stage.

pappus. the modified calyx of flowers in the aster, or sunflower, family; usually takes the form of bristles, scales, or awns.

parasite. an organism that lives in or on the body of another organism (the host) from which it derives its food without killing the host directly; also used in this book to describe an insect that spends its immature stages in the body of a host, which is killed just before the parasite pupates.

pathogen. a disease-causing organism.

perennial. a plant that can live 3 or more years and flower at least twice.

periderm. several layers of corky cells located on the outside of the epidermis of a potato tuber and containing high amounts of suberin.

persistent virus. a virus that systemically infects its insect vector and is usually transmitted for the remainder of the vector's life; circulative virus.

petiole. the stalk connecting a leaf or leaflets to a stem (see Fig. 21).

pheromone. a substance secreted by an organism to affect the behavior or development of others of the same species.

phloem. the food-conducting tissue of a plant vascular system (see Fig. 5).

photosynthesis. the process whereby plants use light energy to form sugars and other compounds needed to support growth and development.

physiological disorder. a disorder caused by factors other than a pathogen; abiotic disorder.

phytoplasma. a microorganism, smaller than a bacterium and lacking a rigid cell wall, that infects plants and is spread by insect vectors; cannot grow or reproduce outside of host plant or insect vector.

phytotoxicity. the ability of a material such as a pesticide or fertilizer to cause injury to plants.

PLRV. *Potato leafroll virus.*

postemergence herbicide. in this book, an herbicide applied after potatoes emerge; also, an herbicide applied after target weeds emerge.

predator. an animal that attacks and feeds on other animals (the prey), usually eating most or all of the prey and consuming many prey during its lifetime.

preemergence herbicide. in this book, an herbicide applied before potatoes emerge; also, an herbicide applied before target weeds emerge.

primary bloom. the first production of flowers on a potato plant, occurring after 8 to 12 leaves have been formed on the main stem and generally coinciding with the beginning of the tuber growth phase.

primary inoculum. the initial source of a pathogen that starts disease development in a given location.

propagule. any part of a plant from which a new plant can grow, including seeds, bulbs, rootstocks, etc. Also, any structure of a pathogen that can serve as a source of inoculum.

protectant fungicide. a fungicide that protects a plant from infection by a pathogen.

pupa (plural, pupae). the nonfeeding inactive stage of an insect between larva and adult in insects with complete metamorphosis.

PVA. *Potato virus A.*

PVS. *Potato virus S.*

PVX. *Potato virus X.*

PVY. *Potato virus Y.*

reservoir. the site where a pest population or quantity of inoculum can survive in the absence of a host crop, and from which a new crop may be invaded.

resistant. able to withstand conditions harmful to other strains of the same species.

rhizome. a horizontal underground stem, especially one that roots at the nodes to produce new plants.

rogue. to remove diseased plants from a field.

rolling. mechanical crushing of potato vines to hasten vine death, sometimes used synonymously with vine killing.

rootstock. an underground stem or rhizome

rosetting. abnormal growth in which new potato foliage is stunted and tightly bunched, caused by certain pathogens.

rugose. a rough appearance of leaves in which veins are sunken and interveinal tissue raised, caused by certain virus infections.

russeting. thickening of the periderm on tubers of russet cultivars that occurs after vine senescence.

sclerotium (plural, sclerotia). a compact mass of hardened mycelium that serves as a dormant stage for some fungi.

secondary bloom. a second production of flowers on a potato plant, occurring at the end of the main stem of an indeterminate cultivar; secondary bloom may occur on a determinate cultivar at leaf axils along the main stem.

secondary infection. infection by a pathogen that enters the host through an injury caused by another pathogen.

secondary pest outbreak. the sudden increase of a pest population that is normally at low or nondamaging levels caused by the destruction of natural enemies by treatment with a nonselective pesticide to control a primary pest.

secondary spread. the spread of a pathogen within a field after the initial or primary infection.

secondary stems. stems formed by stolons that emerge from the soil.

seed leaf. leaf formed within a seed and present on a seedling at germination; cotyledon.

seed piece. portion of a potato tuber containing at least one eye that is planted to produce a new potato plant.

senescent. growing old; aging.

sheath. the part of a grass leaf that encloses the stem below the collar region (see Fig. 40).

soil profile. a vertical section of the soil through all its horizontal layers, extending into the parent material.

specific gravity. the ratio of the density of a substance to the density of pure water; specific gravity of potato tubers is used as a measure of their dry matter content.

sporangium (plural, sporangia). a structure containing asexual spores.

spraing (sprain). reddish brown spots, rings, or arcs in tuber tissue caused by TRV; corky ringspot.

sprout. the new stem formed from the eye of a potato tuber.

sprout inhibitor. a chemical applied to potato vines or to stored tubers to prevent sprouting.

stolon. the underground stem of a plant, the end of which may form a tuber; stolons of some plants often root at the nodes to form new plants.

stoma (plural, stomata). natural opening in a leaf surface that allows gas exchange and water evaporation and has the ability to open and close in response to environmental conditions.

stroma. a compact, usually spore-producing structure formed from fungal mycelium on the surface of a host.

suberin. a waxy substance, resistant to microbial attack, formed in the corky cells of periderm layers.

suberization. the formation of periderm layers on the cut surfaces or wounds of potato tubers.

submarine tubers. small tubers formed on stolons that emerge directly from the eyes of seed tubers

systemic. capable of moving throughout a plant or other organism, usually in the vascular system.

tensiometer. a device used for measuring soil moisture that consists of a closed, buried tube that develops a partial vacuum as the soil dries.

tertiary bloom. the third production of flowers that occurs at the end of the growing stem of an indeterminate potato cultivar.

tolerance level. maximum percentage of a disease or pest symptom allowed during field inspections for certification of a seed lot; levels are different with each field generation and may vary from state so state.

tolerant. a cultivar that is able to grow and produce an acceptable yield when infected by a pathogen.

toxin. a poisonous substance produced by a living organism.

translocated herbicide. the preferred term for a systemic herbicide.

transpiration. the evaporation of water from plant tissue, mostly through stomata.

treatment threshold. the level of a pest population, usually measured by a specified monitoring method, at which a control measure is needed to prevent eventual economic injury to the crop.

true leaf. any leaf produced after the seed leaves (cotyledons).

TRV. *Tobacco rattle virus.*

tuber. an enlarged, fleshy, underground stem with buds capable of producing new plants.

tuberization. the formation of tubers at the ends of stolons; tuber initiation.

vascular ring. a thin area of potato tuber tissue between the cortex and the medulla in which vascular tissue is concentrated (see Fig. 6).

vascular system. the system of plant tissues that carries water, mineral nutrients, and products of photosynthesis throughout the plant, consisting of xylem and phloem (see Fig. 5).

vector. an organism capable of transmitting a pathogen to a host.

vegetative growth. growth of leaves, roots, and stems, including tubers, as opposed to flowers or fruits.

vein banding. dark brown discoloration of the veins on the undersides of potato leaflets caused by PVY.

virulent. capable of causing a severe disease; strongly pathogenic.

virus. a small infectious agent, consisting of nucleic acid and a protein coat, that can reproduce only within the living cells of a host.

wound healing. suberization.

xylem. tissue that conducts water and mineral nutrients from the roots to the rest of the plant (see Fig. 5).

zoospore. a motile spore.

Index